AF323229

Geospatial Computing in Mobile Devices

Geospatial Computing in Mobile Devices

Ruizhi Chen

Robert E. Guinness

ARTECH HOUSE

BOSTON | LONDON

artechhouse.com

Library of Congress Cataloging-in-Publication Data
A catalog record for this book is available from the U.S. Library of Congress.

British Library Cataloguing in Publication Data
A catalogue record for this book is available from the British Library.

Cover design by Igor Valdman

ISBN 13: 978-1-60807-565-2

© 2014 ARTECH HOUSE
685 Canton Street
Norwood, MA 02062

10 9 8 7 6 5 4 3 2 1

Contents

Preface

When we started the project of writing this book more than two years ago, we knew it would be challenging for the following reasons:

- The relevant topics encompassed by "mobile geospatial computing" are both large in number and diverse in nature.
- As the technologies surrounding mobile devices continue to develop at a rapid pace, the state-of-the-art in mobile geospatial computing advances at an equally fierce pace.

Nonetheless, we knew it would be a worthwhile endeavor, considering the great number of people interested in these topics and the many exciting innovations made possible using geospatial computing in modern mobile devices. Furthermore, we knew there was no other book on the market covering this range of topics.

Specifically, the aims of this book are:

- To provide a comprehensive introduction to the field of mobile geospatial computing with particular emphasis on mobile positioning, mobile GIS, location-based services, and context awareness;
- To equip readers with the intellectual basis necessary to contribute to the field of mobile geospatial computing;
- To review the historical development of key technologies and concepts central to mobile geospatial computing.

The target audience of the book includes the following groups:

- Upper-level undergraduate students and master's/doctoral students in fields related to mobile geospatial computing;
- Scientists and engineers working in disciplines where some knowledge of mobile geospatial computing is required;

- Mobile developers and researchers who desire a deeper understanding of geospatial computing topics;
- Business developers, innovators, and entrepreneurs who wish to enhance their technical background related to mobile geospatial computing, especially concerning the current challenges and opportunities in this fast-moving field.

The level of prerequisite knowledge necessary to understand this book is relatively low. Some chapters (Chapters 2–5 and Chapter 9) assume a basic understanding of matrix algebra, calculus, and probability, but the core concepts are described in such a way that those who wish to skip the mathematical details may do so without losing the main thread of discussion. A few chapters assume basic familiarity with computer programming, specifically Chapters 6 and 8, but again the key topics are understandable even for those who have no programming experience.

For those wishing to go deeper into the programming aspects of mobile geospatial computing, we will maintain a website containing example programs covering many of the concepts addressed in this book. These have been written for the Android platform, as we have found this to be the most accessible mobile operating system, especially for beginners. For those teaching courses related to this book, we will prepare lecture materials, including many of the figures and equations appearing in the book. We also welcome feedback via the website on areas of this book that can be further developed in the future, as well as further code examples that readers may wish to see. Lastly, we have worked hard to ensure that no errors appear in the book, but nonetheless it is rare for a book of this nature to go to print without a few stubborn oversights. We would be grateful if readers find these and bring them to our attention, so they can be corrected in future printings. Visit the book's website for further information, which you can find from the link: http://bit.ly/geospatialcomputing.

We hope your experience reading this book will be as enjoyable and rewarding as it has been for us to write it. Most importantly, we hope it inspires and motivates many people to either join or further contribute to this exciting field.

Acknowledgments

We would like to thank the entire team at Artech House who has contributed to the production of this book. Especially, we thank Aileen Storry for her careful and patient guidance throughout her involvement in this project. We also heartily acknowledge the work of Artech House's anonymous reviewer, who provided a great deal of useful feedback, which helped us to improve the text. We would also like to thank Dr. Xiang Li from East China Normal University for his help in the early phase of this project. In addition, we thank Robert J. Guinness, who provided valuable advice on legal matters during the early phase of the project. Lastly, Dr. Joseph Guinness of North Carolina State University provided helpful review and feedback concerning several chapters, which is greatly appreciated.

From R.C.:

During this project, I changed my job from Finnish Geodetic Institute to Texas A&M University Corpus Christi. My colleagues from both organizations are highly appreciated for their support of my research work in this field, which is important to this book. I would also like to thank my wife and daughter for their understanding and support of this book project.

From R.E.G.:

First I would like to acknowledge my friends and colleagues at the Finnish Geodetic Institute with whom I have the pleasure of working. I have learned a great deal from you and have benefited from a collegial work environment. Next I would like to thank my family for your loving support, without which this book would not be possible. I especially appreciate your patience and understanding during a busy phase of editing that took place during Christmas 2013. With this book project behind me, I look forward to spending more time on three very important little "projects" named *Pyry, Kilian,* and *Lila*.

Chapter 1

Introduction

In case you haven't noticed, the technological world is undergoing a revolution—and not a small one. People are walking around with powerful mobile devices in their pockets, handbags, and backpacks. Increasingly, you can even find people walking around while carrying these devices in their hands and staring into their tiny screens, so beware!

This chapter introduces the topic of geospatial computing in mobile devices. The terms *geospatial computing* and *mobile devices* have been chosen carefully, and we hope the reasons will shortly become clear. The chapter's aim is to motivate the remainder of the book by taking a look at how and why these two topics have become so important during recent years and how it has become possible to perform geospatial computing in mobile devices. Section 1.1 explores in more detail the technological revolution mentioned above, which we call the *mobile revolution*. Section 1.2 provides an overview of the field of geospatial computing. Section 1.3 gives an overview of mobile device positioning, which is an important aspect of geospatial computing and one of the focuses of this book. Finally, the overall organization of this book is presented in Section 1.4.

1.1 THE MOBILE REVOLUTION

The mobile revolution started in the so-called developed world—North America, Europe, Australia, and Japan to name a few places—sometime around the mid to late 1990s. In fact, it is difficult to pinpoint an exact year when it began, but certainly by the turn of the millennium, the revolution was in full swing in many parts of the world. Increasingly it has been spreading to everywhere that there are people who can afford these mobile devices. Figure 1.1 provides a glimpse into how this revolution has played out in a number of parts of the world. As the prices of mobile devices continue to drop (per unit of computing power), more and more people are taking them into use. Furthermore, those that have already used mobile devices for many years are continuously upgrading their old devices to newer and more powerful ones. As a result of these trends, mobile devices are quickly being integrated into our routines of daily life—not just in a few countries, but in practically all countries.[1]

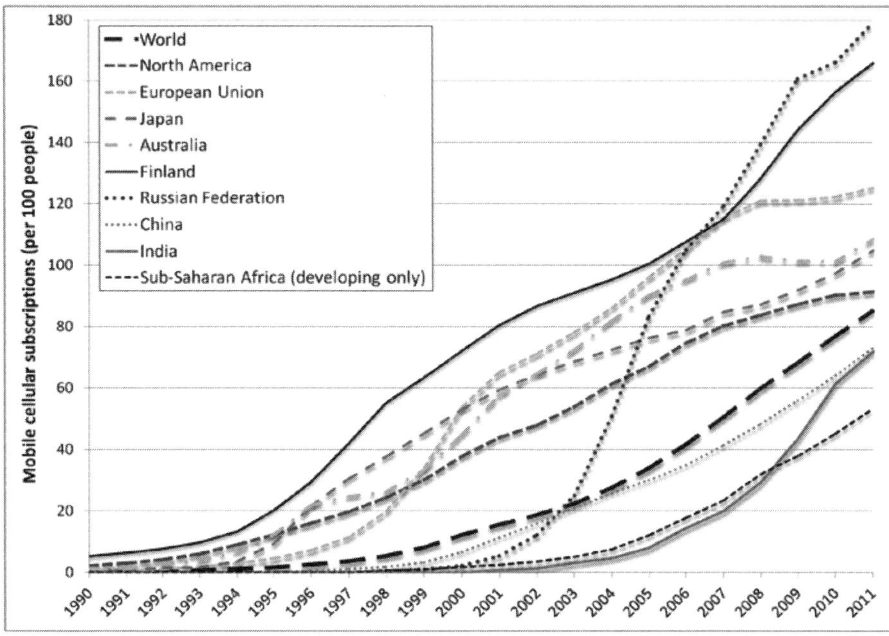

Figure 1.1 Growth of mobile subscribers in different parts of the world (Data source: World Bank) [1].

1.1.1 Terminology of the Mobile Revolution

At various points in their developmental history, mobile phones went by other names (e.g., car phones and cell phones). At some point after mobile phones began incorporating advanced capabilities, such as personal digital assistant (PDA) applications, data connectivity, and positioning capability, people began referring to these high-end phones as "smart phones" (which eventually morphed into "smartphones"). There is no clear division between which phones are considered "smart" and which ones are not. At the low end of the cost spectrum, marketers call modern mobile phones *feature phones*, since they incorporate a sizeable subset of the features from bona fide smartphones. One reason that there

---

[1] A few notable exceptions are North Korea, Somalia, and Eritrea, where use of mobile devices is severely lagging behind the rest of the world [1].

is no clear definition of a smartphone is because mobile technology is developing rapidly. To put it simply, today's feature phones were yesterday's smartphones!

We should also mention that today's mobile devices come in many shapes and sizes (albeit mostly all rectangular shapes). At the big end of the size spectrum, we call these mobile devices *tablets*. You would have trouble fitting these in your pocket, but they certainly can fit in a small backpack. Although it is possible to place phone calls on many tablets, this is not their primary purpose. This is one reason we have chosen the term *mobile devices* in this book, instead of focusing solely on mobile phones.

Smartphones, feature phones, and tablets are not the only mobile devices relevant to this book. Marketers are now even referring to medium-sized mobile devices as *phablets*. Furthermore, there is a newly arriving class of miniature mobile devices known as wearable devices. This includes such things as *smartwatches* and *smartglasses* (e.g., Google Glass). For the purposes of this book, all these devices—from tablets to wearable devices—can be considered *smart mobile devices*.

In practical terms, much of the material in this book applies to all classes of smart mobile devices. They differ only in their screen size, computing power, battery size, and other basic specifications. Otherwise, they operate in essentially the same ways. The one group of mobile devices that may not be relevant to this book is low-end feature phones. For example, if a feature phone cannot easily run third-party applications or does not have positioning capability, there is little hope that one will be able to apply many of the concepts from this book using such a phone. For roughly the same price as this book, however, you can already buy a decent mobile device. We highly recommend that readers get practical experience with at least one such device, in order to better appreciate the topics in this book!

1.1.2 Challenges and Opportunities of the Mobile Revolution

As people begin to walk around with these powerful mobile devices—in their pockets, handbags, backpacks, or even in their hands—we are presented with both challenges and opportunities. The challenges are primarily the following:

- The computing resources of mobile devices are relatively limited compared to personal computers (PCs), such as laptops or desktop computers. In particular, they are mostly used as battery-operated devices, so careful attention must be paid to minimize energy usage.
- Compared to PCs, these devices are highly portable. Thus, they will be used in many environments where PCs are not commonly used, such as in cars, buses, and the outdoors. As a result, special care must be taken to ensure good usability in a wide range of situations and environments. One related challenge is that positioning systems developed prior to the mobile revolution (namely GPS and other satellite-based navigation systems) are not designed

for indoor usage. In fact, Chapters 4 and 5 are mainly concerned with addressing this issue.

On the other hand, the opportunities offered by smart mobile devices are many. Here we highlight a few unique opportunities they present for developers, scientists, engineers, and business people:

- As a corollary to the second item above, because these devices are highly portable, people keep them with them most of the time. Thus, there is an opportunity to use them in new ways that weren't possible with PCs and other earlier computing devices. Furthermore, applications on these devices can exploit knowledge of the users' positions and movements. For example, 24/7 tracking applications are to a large extent now feasible.
- Because of the decreasing cost and increasing utility of smart mobile devices, in many countries most people have at least one. Thus, the market potential for new applications is enormous. In many ways, it is larger than for traditional PC software. In addition to business considerations, smart mobile devices can be used by scientists in new ways to accomplish what may have previously seemed infeasible. For example, sociologists could potentially use mobile devices to study how people move en masse. By employing crowdsourcing, a wide variety of data can be collected for a very low cost. This is not just relevant for scientists, but also for business people and application developers who wish to make use of such data in novel ways.
- The third opportunity is less concrete but equally important. It results from the fact that smart mobile devices are changing people's lifestyles and expectations in many ways. One could argue that people are becoming more mobile *as a result* of smart devices. For example, many people spend significant amounts of their working time on the road or in "satellite" locations (e.g., in a coffeehouse). In many cases, this working style would not be possible—or at least not as feasible—without mobile devices. People's expectations about what should be possible with smart mobile devices and geospatial computing are also increasing. For example, "Why can't I control my house locks or garage door with my smartphone?" These kinds of expectations present many new opportunities for geospatial computing applications.

Equipping our readers with the tools to address these challenges and seize these opportunities will be the primary goal of this book. By the end of the book, we hope you will be ready, if you aren't already, to make your first location-aware mobile application, perhaps named *GarageOpenerApp*!

1.2 INTRODUCTION TO GEOSPATIAL COMPUTING

This section introduces the topic of geospatial computing in a general sense. We will first define the term, as well as discuss several related disciplines. Next, we will look at recent developments that have brought about what some have called the *geospatial era*. Finally, we will briefly discuss some of the important concepts and tasks in geospatial computing.

1.2.1 What Is Geospatial Computing?

A question you may have asked when you first saw the cover of this book is, "What exactly is geospatial computing?" This is understandable given the fact that it is a relatively new term. In fact, if you type "geospatial computing" (including quotes) into Google, you're likely to get a modest 5,000 to 10,000 results.[2] The somewhat broader term, "spatial computing," will yield 25,000 to 30,000 results. Neither term can be found in any of the major dictionaries, such as *Merriam-Webster* or *Oxford English*. These facts suggest both are not yet well-established terms.

Geospatial computing is a set of computational tasks that involve geospatial data or information. The terms *geospatial data* or *geospatial information* refer to data/information that is spatially referenced with respect to Earth. Some may wonder what the difference is between the terms *geospatial* and *geographic*. They are, in fact, synonyms, but we tend to view geospatial as encompassing a wider gamut of entities and abstract concepts. The adjective *geographic* tends to apply to relatively static entities on the Earth, such as landforms, bodies of water, cities, and property boundaries. Geospatial information, on the other hand, can refer to any type of information that is spatially referenced with respect to Earth. The term "geospatial" itself dates back to at least the early 1970s (see [2]).

Perhaps this difference is best elaborated with an example: The trajectory of a motor vehicle (on Earth) would not generally be considered "geographic information" since it is not concerned with a static (or lasting) feature of the Earth. It would most certainly, however, be considered geospatial information because any meaningful trajectory would be spatially referenced (i.e., expressed in some kind of geographic coordinate system).

One of the first uses of the term geospatial computing was in an article by Egenhofer, who described a vision of "ubiquitous geospatial computing" dominated by "spatial information appliances" [3]. Today this vision bears a striking resemblance to the present day ubiquity of smart location-aware devices, such as those described above.

---

[2] Note that the exact results of such queries will vary from country to country and user to user. Also, note that without putting the phrases in quotes, Google will return a much bigger number of results.

1.2.2 Related Disciplines

There are many "cousins" to geospatial computing that deserve mention. For example, geospatial computing is closely linked to geographic information science [or similarly geographic information systems (GIS)], but it encompasses a somewhat wider range of subjects. Geospatial computing is more or less synonymous with the term *geoinformatics*, but we prefer the former for this book because our focus is on all computational issues related to geospatial data/information, rather than those typically associated with informatics.

Another important related field is *computational geometry*, which is a branch of computer science that tries to find efficient algorithms and data structures to solve geometric problems. One simple example would be, given a set of points S, find the point in S nearest to point q. Since geospatial computing frequently requires solving geometric problems, computational geometry is an important subject in geospatial computing. In some sense, the whole field of geospatial computing can be considered a subset of problems in computational geometry.

With regards to generating geospatial data, an important discipline is *remote sensing*, which, as the name implies, deals with acquiring data at a distance using various sensors. Most remote sensing data comes from sensors flown on aircraft or spacecraft. Typical sensors include imaging sensors (visible, infrared, or other portions of the electromagnetic spectrum), radar sensors, or ranging sensors [such as light detection and ranging (LiDAR)]. Remote sensing can be employed to perform regional or global measurements of various kinds, such as measurements of the Earth's gravitational field, oceanic and atmospheric measurements, and environmental change and land usage, or even to detect usage of nuclear weapons. We will not cover remote sensing in much detail in this book, but it is important to have basic knowledge and understanding of how the geospatial data you are using was acquired.

Last, an important subject related to geospatial computing is *geodesy*. Geodesy is a branch of applied mathematics dealing with mathematical representations of the Earth, including its position and orientation in space, as well as the science of measuring the size and shape of the Earth. Geodesy helps to define *geographic coordinate systems*, which will be described in more detail below. *Geodetic surveying* refers to methods of measuring various properties of the Earth, such as its gravitational field, magnetic field, or crustal movement, at one or more points in space. In such surveys, great care is taken to accurately reference the measurements within a specific geographic coordinate system. Thus, methods of accurate positioning are highly important in geodesy.

1.2.3 The Geospatial Computing Era

Despite the fact that the term geospatial computing was not used until the 1990s, the concepts and technologies related to geospatial computing are much older. In particular, development of the first GIS began in the 1960s with work by

Tomlinson to create the Canada Land Inventory (CLI) [4]. By the early 1990s, there were a large number of publications related to GIS and other geospatial technologies. For a more detailed history of GIS developments, see [5].

According to Gary Hacker of the National Imagery and Mapping College in the United States, the use of such geospatial technologies, especially for the process of mapmaking, became widespread by 1993:

The geospatial era, which began in 1993, is most notably marked by the near simultaneous introduction in the public sector of three technologies heretofore virtually reserved for government use: high-resolution digital satellite remote sensing imagery (RSI), geographic information systems (GIS) and global positioning system (GPS) [6].

One source of evidence to support this claim is the geospatial industry that sprung up around this time. For example, the leading GIS company, ESRI, Inc., had $120 million in sales in 1993 [7]. In 1995, the U.S. GPS Industry Council estimated the total GPS market to be worth nearly $1.3 billion [8]. By 1996, the GIS industry alone was worth $850 million worldwide [9]. As discussed earlier and cited in the above quote, remote sensing is an important discipline related to geospatial computing. For a detailed historical treatment of RSI, we refer the interested reader to [10].

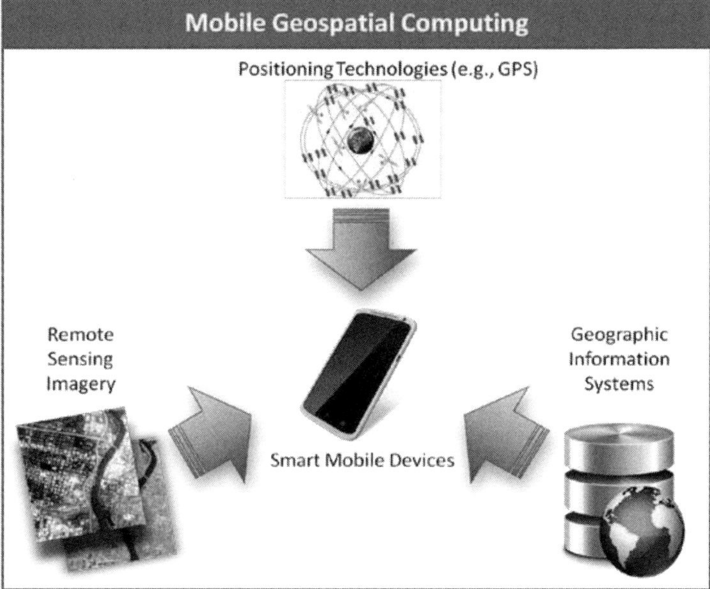

Figure 1.2 Mobile geospatial computing is a convergence of three major technologies into the smart mobile device platform: positioning technologies, GIS, and RSI.

In addition to commercial activities, governments continued to develop technologies related to geospatial computing during the nascent geospatial era. For example, in the early 1990s the United States began discussions about establishing a National Spatial Data Infrastructure (NSDI) [11]. This was formally established by former President Clinton in April 1994 to include "the technology, policies, standards, and human resources necessary to acquire, process, store, distribute, and improve utilization of geospatial data" [12]. Nancy Tosta of the U.S. Federal Geographic Data Committee noted in 1994 that "the feasibility of developing the NSDI has increased steadily over the last several years as geospatial data computing and telecommunications networks have become more pervasive [13]."

Putting these developments together, we can see that the mid1990s saw a confluence of technologies, commercial actors, and government policies that marked the beginning of the so-called *geospatial era*. Combined with the fast development of smart mobile devices described in Section 1.1, the geospatial era is quickly morphing into the mobile geospatial computing era, which this book aims to address. This is visualized in Figure 1.2.

1.2.4 Important Concepts and Tasks in Geospatial Computing

In this section we will briefly discuss some of the important concepts and tasks in geospatial computing. Some (but not all) of these topics will be covered in greater depth in later chapters.

Coordinate systems and projections: Since geospatial computing deals with geospatial information, the first thing we require is an understanding of coordinate systems and projections. A *coordinate system* is a numeric model of space, which can be of arbitrary dimension. Typically in geospatial computing we work with two- or three-dimensional coordinate systems.

The most commonly used coordinate systems in geospatial computing are called *geographic coordinate systems*, which are based on the surface of the Earth. One of the tasks of geodesy is to derive a model of this surface, called the geoid, which is typically in the form of an ellipsoid. *Latitude, longitude*, and *height* (e.g., above the geoid) are the typical dimensions of geographic coordinate systems. Cartesian coordinates can also be used, where the surface of the Earth over a small area is approximated as a plane. This raises the topic of *projections*, which are transformations between different kinds of coordinate systems, such as projections of a spherical surface onto a plane. These topics will be discussed further in Chapter 2.

Positioning: Positioning is the process of determining the coordinates of an object or person within a coordinate system. The choice of coordinate system is entirely dependent on the application, but most often a geographic coordinate system is used. Note that positioning can be considered one of many subtasks in *navigation,* which may also include determining one's velocity and heading, as well as determining the best route to follow towards one's destination. Positioning will be discussed further in Chapters 2–5, and Chapter 9 will discuss a few other aspects of navigation.

Storage of geospatial data: Just as there are many different types of geospatial data, there are many different ways to store such data. Two major types of geospatial data, especially with regard to geospatial images, are *vector data* and *raster data*. Vector data consisted of points, lines, and polygons, where the coordinates of each point or line segment are specified in a particular coordinate system. Vector data are typically stored in a file or database for use by GIS or other software applications. Raster data consist of an evenly spaced grid of pixels, where each stores a single value. These data are typically stored in an image file format, but they can also be stored in a database. Raster data can easily be *georeferenced* by storing the coordinates of two or more pixels (or alternatively by storing the coordinates of, for example, the upper left pixel, as well as the spacing and orientation of data). In any type of geospatial data, it may be desirable to store a *spatial index* created from the data itself. A *spatial index* is a special type of database index that is optimized for performing spatial queries.

Geospatial analysis: Given a geospatial dataset, one important task is *geospatial analysis*, which can be defined as the extraction of high-level information from low-level geospatial data. This is one area where computational geometry is heavily applied. In addition to purely geometric computations, geospatial analysis includes such tasks as building statistical models from geospatial data, performing classification of data entities based on geospatial properties, or other forms of descriptive or exploratory analysis, such as finding clusters of geospatial entities from a large dataset.

Geospatial visualization: Another important task in geospatial computing is to visualize geospatial data, in order to clearly convey some particular information, such as the spatial distribution of a particular group of entities. *Cartography* is a discipline that focuses heavily on geospatial visualization, especially with regard to displaying geospatial information on a map. The cartographer's task is to convey such information as clearly but accurately as possible, which often requires approximating complex geospatial entities using simpler ones. For example, a long river could have a complex meandering shape. A visualization task may be to plot this river as a polygon at various scales and with varying levels of complexity. Accomplishing visualization requires good use of computational geometry, but often it also requires creativity and design skills to determine the best way to present information.

Image processing and computer vision: The above example raises the topic of *image processing*, which is a common task involved in geospatial computing. As mentioned above, remote sensing provides an important source of geospatial data, and much of it comes in the form of images. Therefore, it should not come as a surprise that one needs to understand how to work with images and process them to extract various kinds of information. A case in point is obtaining the initial shape of the meandering river from our above example, which is likely to be performed by processing a high-resolution remote sensing image.

The set of methods to process images and extract high-level information is often known as *computer vision*. The name comes from the fact that computer vision tasks often involve mimicking human vision using computers. For example, a meticulous person might be able to count all of the buildings in an aerial image, but writing a computer program to do so reliably is not a trivial task.

We hope that this section has given the reader a sense of the breadth of subjects and the exciting challenges that lie ahead in the field of geospatial computing. Obviously a single book cannot cover all of these subjects in great detail. We will focus on a subset of these subjects, focusing on those that are most relevant to geospatial computing in mobile devices.

1.3 INTRODUCTION TO MOBILE DEVICE POSITIONING

For hundreds, if not thousands of years, scientists and engineers have tried to come up with new and better devices and techniques for positioning and navigation. For example, the *compass* has been used for navigation since at least 1088, when it was described by Chinese polymath Shen Kuo [14].[3] Another device, the mariner's astrolabe, used the position of stars (including the Sun) for determining the latitude of a ship at sea. The network of thousands of lighthouses around the world (beginning with Pharos of Alexandria built by the Egyptians in 280 BC) was also built up primarily for the purposes of navigation. Thus, we can see that the science of navigation and positioning has a long tradition. Modern devices, networks, and signals that we use for navigation are clear descendents of these ancient ones.

As far as mobile device positioning, there are primarily three techniques in use today:

- Those using global navigation satellite systems (GNSS);
- Those using cellular networks (namely mobile phone networks);
- Those using wireless local access networks (WLANs).

---

[3] The compass was actually invented much earlier in China, but it was not initially used for navigation purposes.

The technical aspects of all three of these techniques will be described in detail in Chapters 2–5. In the sections below, we will present a historical overview of how these different techniques came into being.

Since the term GNSS has not yet been discussed, let us briefly define this concept for readers who are unfamiliar with the term. A GNSS is any satellite system providing global navigation services (i.e., position, velocity, and time solutions), for end users. The GPS is the most widely known GNSS, but since this name refers specifically to the U.S. system, it is not correct to call, for example, the Russian system, a GPS. Many commercial receivers now use signals from multiple GNSSs, so it is more accurate to refer to these as GNSS receivers, rather than GPS receivers. The navigation community has long ago adopted this more general terminology, and it is likely that the mobile industry will soon follow suit.

1.3.1 Predecessors to Global Satellite Navigation Systems

In the early twentieth century scientists and engineers began experimenting with the use of radio signals for navigation. By World War II, various radio navigation techniques were in widespread use by both the Allied and Axis powers, such as radio beam navigation (i.e., A-N system), the British Gee system, radio direction finding, and LORAN [15].[4]

After WWII, radio navigation using ground-based radio beacons continued to develop further (e.g., VOR, LORAN-C, and OMEGA), but scientists, such as Yrjö Väisälä, also began to use high-altitude weather balloons and triangulation to determine positions of points on Earth, in order to establish more accurate geodetic reference frames [16]. Instead of radio signals, visible light signals were flashed from the balloons and photographed against a backdrop of stars. In the late 1950s, after the dawn of the space age, the United States realized it could use Earth-orbiting satellites to achieve even greater accuracy and global coverage. It began developing the TRANSIT satellite navigation system, as well as additional satellites used for geodetic positioning (e.g., Echo I, PAGEOS, and SECOR satellites) [17, 18]. In addition, the Soviet Union began developing the Tsiklon, Tsikada, and Parus satellite systems, which operated similar to the TRANSIT system [19].

Around the same time that TRANSIT became operational (circa 1963–1964), there were already several programs initiated to build a more capable successor. The U.S. Navy Research Laboratory (NRL) began developing the Timation system, and the U.S. Air Force (USAF) funded System 621B, developed by the Aerospace Corporation. These efforts were to be combined after the Department of Defense (DoD) established the Navigation Satellite Executive Group (NAVSEG) in 1968 to consolidate efforts in satellite navigation [20].

---

[4] Officially LORAN is an acronym for long-range navigation, but its original name was Loomis radio navigation (LRN), after its inventor Alfred L. Loomis.

1.3.2 The Beginning of the GNSS Era

In April, 1973, the DoD directed the Air Force to begin developing the defense navigation satellite system (DNSS), which later became known as GPS [8]. The successor to the Soviet satellite-based navigation systems is the Globalnaya Navigatsionnaya Sputnikovaya Sistema (GLONASS)[5] whose development formally began in 1976 (although technical development likely began much earlier) [20]. Although these systems were developed primarily for military purposes, early in their developmental history the United States and the Soviet Union made some of the satellite signals available for civilian use. Already in April 1981, it was publicly announced that the coarse acquisition (C/A) signal of GPS would be available for civilian use after GPS became fully operational [8].[6]

When these systems became fully operational in the mid1990s, it represented a new era in global, real-time positioning capability. Despite the reservation of the highest accuracy signals for military use, the openly available civilian signals continue to provide unprecedented, worldwide positioning accuracy for the public. For example, after selective availability (SA) was disabled in May 2000, typical horizontal error levels were reduced to around 20m. Such capability is something that navigators centuries ago probably could have never imagined!

It was not long after GPS became fully operational that companies began developing mobile systems that were capable of using GPS for positioning. One example is NAVSYS, which in 1995 developed a prototype emergency response system consisting of a Nokia mobile phone connected to a unit called LocatorNET, which collected raw GPS measurements [21]. These measurements could then be forwarded to a server to calculate the user's position. Another example is Benefon (now Twig Com Ltd.), which began research concerning GPS safety phones already in 1995 and released its first GPS-enabled mobile phone (named *Benefon Esc!*) in 1999 [22].

1.3.3 The E-911 Initiative

In terms of positioning in mobile phones, in many ways it is not GNSS positioning that led the way but rather cellular-based positioning. A major catalyst for this was the adoption of Enhanced 911 (or E-911) regulations in the United States, starting in 1996. These regulations require wireless network operators to be able to locate their users in the case of emergency calls. Circa 1996, GNSS receivers were bulky and power-hungry. For a car-based phone like NAVSYS, this was not a major issue, but the mobile industry was pushing forward toward

---

[5] This is normally translated into English as "Global Navigation Satellite System," which may cause confusion with the general term GNSS. In practice, the acronym GLONASS avoids this problem.
[6] For comparison, the Soviet Union declared in 1991 that GLONASS would be available to the international aviation community, but due to the Soviet Union's collapse, GLONASS remained "incomplete" until 2011 [24].

smaller mobile phones that could be used anywhere. Therefore, they had to develop positioning technology that did not require significant additional hardware or huge power levels. This led to the development of cellular-based positioning techniques.

The E-911 regulations were actually implemented in two phases. In the first phase, which was enforced starting in April 1998, network operators had to report "the location of the base station or cell site receiving a 911 call to the [Public Safety Answering Point]" [23]. The second phase, which required greater location accuracy, was more complicated because the requirements depended on whether providers opted to use network-based techniques (e.g., TOA and AOA) or handset-based techniques (i.e., E-OTD and A-GPS). Increasing requirements were phased in between October 2001 and December 2005. They required accuracies ranging from 50m (67% of calls) to 300m (95% of calls), depending on the positioning method employed [25]. It should be noted that these requirements apply only to outdoor cases. Even today no location accuracy requirements exist for 911 calls made from a mobile phone located indoors.

Cellular-based positioning remains an important topic today, and it will continue to be important in the future. Positioning procedures have been incorporated into new cellular standards, such as long term evolution (LTE), and new techniques are being actively researched and discussed (e.g., [26]). Thus, despite the rapid progress of GNSS technologies in recent years, cellular-based positioning is in many ways keeping pace by offering moderate positioning accuracy with much lower power requirements compared to GNSS technology.

1.3.4 Using WLANs for Positioning

In the late 1990s, as WLAN technology experienced rapid rates of deployment, some researchers began investigating how these types of signals could be used as well for positioning. In this way, such developments are similar to those of cellular-based positioning in that the used signal is not intended primarily for positioning purposes. In the positioning domain, these are often referred to as "signals of opportunity." In other words, they are there and mobile devices can receive them, so why not try to use them for positioning? One key difference, however, between cellular-based positioning and WLAN-based positioning is that the latter is typically targeted for indoor positioning applications. Indoor environments are where these signals are the strongest, whereas other signals, such as GNSS and cellular signals, are comparatively weak.

One of the first known experiments using IEEE 802.11-compliant WLAN signals was done by Microsoft researchers and published in early 1999. They reported 50th percentile error of around 3m. Within a few years of this pioneering work, many groups around the world began performing related research using WLAN signals. In addition, several companies related to WLAN-based positioning, such as Ekahau, Inc. and SkyHook Wireless, were founded around this time. Today WLAN-based positioning is used by all of the major mobile device operating systems, including Android, iOS, and Windows Phone.

1.4 ORGANIZATION OF THE BOOK

This book is organized into two main sections. Section 1 covers in detail the topic of positioning in mobile devices, whereas Section 2 addresses other important aspects of geospatial computing, including its applications. The chapters can be read in any order with the exception that Chapter 7 should be read before Chapter 8.

The following is a brief overview of the remaining individual chapters. Chapter 2 addresses the fundamental concepts and issues related to mobile positioning. Chapter 3 concentrates on GNSS-based positioning, whereas Chapter 4 covers mobile positioning based on other wireless communications signals (such as cellular networks and WLAN). Chapter 5 describes hybrid positioning, which involves the use of more than one positioning method. Chapter 6 covers GIS in mobile devices. Chapters 7 and 8 are dedicated to the topic of context awareness. Chapter 7 can be viewed as an introduction to context awareness (which is why we recommend reading it before Chapter 8), whereas Chapter 8 dives deeper into the topic, describing the algorithms that can be used for inferring the mobile context. Chapter 9 covers applications of mobile geospatial computing, which are typically known as location-based services. Finally, Chapter 10 contemplates where mobile geospatial computing is headed in the future and offers our conclusions concerning this exciting field as a whole.

References

[1]   World Bank Group, "Data: Indicators," 2013, available online: http://data.worldbank.org/indicator.

[2]   Kohn, C. F., "The 1960's: a decade of progress in geographical research and instruction," *Annals of the Association of American Geographers 60*, No. 2 (1970), pp. 211-219.

[3]   Egenhofer, M., "Spatial information appliances: A next generation of geographic information systems," *In 1st Brazilian Workshop on GeoInformatics*, Campinas, Brazil. 1999.

[4]   Coppock, J. T., and D. W. Rhind, "The history of GIS," *Geographical information systems: Principles and applications 1*, No. 1, 1991, pp. 21-43.

[5]   Foresman, T., *The History of Geographic Information Systems: Perspectives from the Pioneers,* Prentice Hall, 1997.

[6]   Hacker, G. A., *Strategic Model for Future Geospatial Education*, Carlisle Barracks, PA: U.S. Army War College, 1998.

[7]   Ouellette, J., "GPS industry prepares for boom," *Physics Today, 48*, No. 12 (1995), pp. 8-10.

[8]   Pace, S., et al., "The global positioning system: assessing national policies," No. RAND-MR-614-OSTP, Rand corp Santa Monica, CA, 1995.

[9]   Washington Technology, "A Picture Is Worth a Billion Dollars," Sep 26, 1996.

[10] Campbell, J. B., and R. H. Wynne, *Introduction to Remote Sensing*, New York: The Guilford Press, 2011.

[11] Federal Geographic Data Committee, "The Federal Geographic Data Committee: Historical Reflections–Future Directions," available from http://www.fgdc.gov/library/whitepapers-reports/white-papers/fgdc-history.

[12] Clinton, W. J., "Executive Order 12906 of April 11, 1994," available from: http://www.archives.gov/federal-register/executive-orders/pdf/12906.pdf.

[13] Tosta, N., "Standards to support the national spatial data infrastructure," *ACM StandardView*, Vol. 2, No. 3, 1994, pp. 143–147.

[14] Breverton, T., *Breverton's Encyclopedia of Inventions: A Compendium of Technological Leaps, Groundbreaking Discoveries and Scientific Breakthroughs That Changed the World*, Quercus, 2012.

[15] Hollmann, M., "History of Radio Flight Navigation Systems," available from: http://www.radarworld.org/flightnav.pdf

[16] Kakkuri, J., and L. Kivioja, "Global positioning: The early Finnish connection," *Physics Today*, Vol. 65, No. 8, 2012, p. 8.

[17] Guier, W. H., and G. C. Weiffenbach, "Genesis of Satellite Navigation," *Johns Hopkins APL Technical Digest*, Vol. 19, No. 1, 1998.

[18] National Oceanic and Atmospheric Administration, *"Entering the Space Age: The Evolution of Satellite Geodesy at the Coast and Geodetic Survey,"* available from: http://celebrating200years.noaa.gov/foundations/satellite_geodesy/.

[19] Wade, M., *Encyclopedia Astronautica*, available from http://www.astronautix.com

[20] Russian Space Systems, *GLONASS*, available from: http://www.spacecorp.ru/en/directions/glonass/

[21] Lacey, N., and M. Cameron, "Mayday in the Rockies: Colorado's GPS-based Emergency Vehicle Location System," *GPS Word,* October 1995, pp 40-44.

[22] "Benefon Esc!,"*GSMarena.com*, available from http://www.gsmarena.com/benefon_esc!-44.php.

[23] Federal Communications Commien, 1996, *Revision of the Commission's Rules to Ensure Compatibility with Enhanced 911 Emergency Calling Systems,* available from http://transition.fcc.gov/Bureaus/Wireless/Orders/1996/fcc96264.txt.

[24] Russian Federation, "Status and Development of GLONASS," 12th Air Navigation Conference, Montréal, 19-30 November 2012.

[25] Federal Communications Commien, 2000, *Revision of the Commission's Rules to Ensure Compatibility with Enhanced 911 Emergency Calling Systems,* available from http://transition.fcc.gov/Bureaus/Common_Carrier/Orders/2000/fcc00326.pdf.

[26] Zhang, T., D. Xiao, J. Cui, and X. Luo, "A novel OTDOA positioning scheme in Heterogeneous LTE-Advanced systems," *In 3rd IEEE international conference on Network Infrastructure and Digital Content (IC-NIDC), 2012*, pp. 106-110, 2013.

Chapter 2

Fundamentals of Mobile Positioning

One of the most basic tasks in geospatial computing is to determine the position of a mobile device. This chapter introduces the fundamentals of positioning in mobile devices. It introduces the topics of coordinate systems, positioning observables, and positioning methods. These topics will be expanded upon in later chapters, especially Chapters 3–5. Many of the topics in this chapter are fundamental in the sense that they are applicable to the general field of navigation and positioning. Soon, however, we will narrow our focus to that of positioning methods that can be used in mobile devices.

A location is typically defined by coordinates in one, two, or three dimensions in a specified coordinate system. It is typically represented mathematically by a scalar number x for the case of one dimension, or a vector of two or three dimensions such as (x, y) or (x, y, z). An example of a one-dimension position is the distance along a set of train tracks from a certain train station. A typical two-dimensional position is a set of planar coordinates, such as horizontal coordinates (consisting of north and east components) in a grid coordinate system, or the latitude and longitude of a location on the surface of the Earth. A three-dimension position offers a complete description of a particular location in space (e.g., a position with a vector of three-dimensional Cartesian geocentric coordinates).

Positioning is the process of determining the coordinates for some entity using different observables that are functions of the coordinates, for example, a distance, which is a simple function of two locations. A positioning observable is a quantity or attribute that can be either directly measured or indirectly derived from a measurable signal.[7] Classical observables used for positioning include range, range difference, azimuth, and various angles. With the development of various sensor technologies, new observables are now adopted for the purpose of positioning as well. These new observables include acceleration, velocity, rate of angle change, magnetic field, proximity of radio signals, received signal strength

---

[7] Note that *observable* refers to the quantity or attribute itself (e.g., distance), whereas the word *measurement* refers to a measured value of an observable (e.g., 8.12 meters).

(RSS), and various image features. Any quantity or attribute that can be used for the purposes of positioning is, by definition, a positioning observable.

2.1 COORDINATE SYSTEMS

The coordinates representing a position are associated always with a coordinate system. For mobile positioning, the most important types of coordinate systems include the Earth-centered inertial (ECI) coordinate system, the Earth-centered Earth-fixed (ECEF) coordinate system, the local geodetic coordinate system, and the height system.

2.1.1 The ECI Coordinate System

The ECI coordinate system is an inertial coordinate system in the sense that it remains fixed relative to the Earth's center but does not rotate along with the Earth's rotation. For all practical purposes, the orientation of the ECI coordinate system remains fixed relative to all the stars in the Milky Way, except for the Sun (known as the *celestial sphere*). This type of coordinate system is of particular relevance to satellites of GNSS because a satellite's motion can be described with Newton's laws in an ECI coordinate system. The ECI coordinate system is a three-dimensional Cartesian coordinate system. As the name implies, its origin is at the Earth's center of mass. The XY-plane is defined by the Earth's equatorial plane. The Z-axis points to the North Pole, and the X-axis points to the vernal equinox in the celestial sphere. The Y-axis is normal to the XZ-plane and pointing in the direction defined by the right-hand rule, as shown in Figure 2.1.

Due to a variety of phenomena such as the gravitational attractions of the Sun and the Moon, the equatorial plane, the location of the North Pole, and the vernal equinox are moving continuously with respect to the celestial sphere. Therefore, the ECI coordinate system has to be defined with respect to the equatorial plane, the location of the North Pole, and the vernal equinox for a particular reference *epoch*, for example 12:00 UTC (Coordinated Universal Time) on January 1, 2000. The ECI coordinate system defined with respect to this epoch is called the J2000 coordinate system, which is commonly used as the inertial coordinate system for orbit determinations of GNSS satellites.

2.1.2 The ECEF Coordinate System

Because an ECI coordinate system does not rotate along with the Earth, the ECI coordinates of a point on the surface of the Earth vary with time due to the Earth's rotation. Therefore, it is not convenient to use an ECI coordinate system to describe positions on the surface of the Earth.

For such positions, it is more appropriate to use an ECEF coordinate system. In such a system the axes rotate along with the Earth so that the ECEF coordinates of points on the Earth's surface are constant (except for changes caused by geodynamic phenomena such as continental drift).

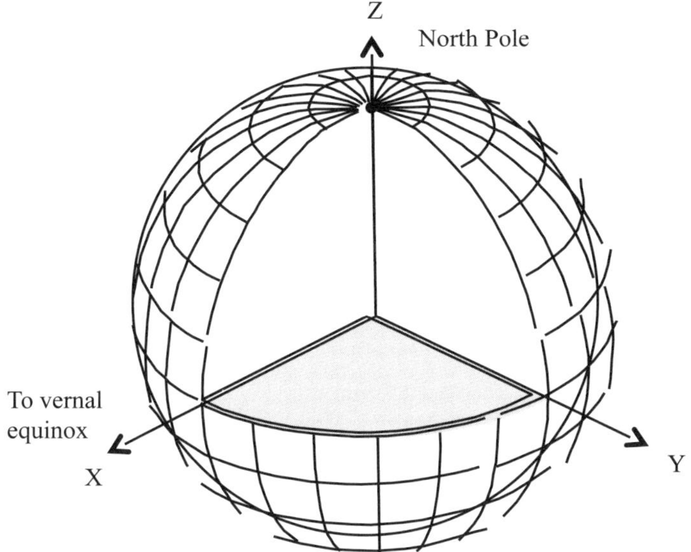

Figure 2.1 The definition of the ECI coordinate system.

Similar to the ECI coordinate system, the ECEF coordinate system is a three-dimensional Cartesian coordinate system with its origin at the Earth's center of mass. The XY-plane of the ECEF coordinate system coincides with the Earth's equatorial plane. The X-axis points toward the intersection point of the equator and the meridian passing through Greenwich, England, also known as the prime meridian. The Z-axis points towards the North Pole, and the Y-axis points to a direction of 90°E to form a right-hand coordinate system, as shown in Figure 2.2. The Cartesian coordinates $[X, Y, Z]$ in an ECEF coordinate system represent a unique location in three-dimensional space; however, people do not generally have a good intuition about locations on Earth according to their Cartesian coordinates. It is much easier for people to understand the geodetic coordinates [latitude, longitude, height]. The definition of a geodetic coordinate system requires the definition of a *reference ellipsoid*, which is designed to follow approximately the shape of the mean sea level around the Earth's surface. The shape of an ellipsoid can be described by the lengths of its semimajor axis a and its semiminor axis b. The cross sections of a reference ellipsoid parallel to its equatorial plane are circles, while those normal to the equatorial plane are ellipses.

For the global coordinate system adopted for GPS, which is called the World Geodetic System 1984 (WGS-84), the length of the semimajor axis a of its reference ellipsoid is 6378.137 kilometers. The length of the semiminor axis b is derived from a defining parameter known as flattening f, according to the following formula:

$$b = a(1 - f) \tag{2.1}$$

The value for reciprocal of flattening $1/f$ chosen for the WGS-84 coordinate system is 298.257223563 [1]. Thus, the length of the semiminor axis b is 6356.7523142 kilometers according to (2.1). Another useful value related to the reference ellipsoid is its eccentricity e, which is related to the semimajor and semiminor axes according to the equation:

$$e = \sqrt{(a^2 - b^2)} / a \tag{2.2}$$

In WGS-84, e has a value of about 0.0818. Note that a sphere would have an eccentricity of zero. By design, the reference ellipsoid describes the "bulging" of the Earth at the equator, due to centrifugal force caused by its rotation about the North Pole.

The definitions of geodetic latitude ϕ and geodetic longitude λ are also depicted in Figure 2.2. The geodetic height h is defined as the normal distance to the surface of the ellipsoid. The geodetic height is positive for a point above the ellipsoid surface, negative for a point below it, and zero for all points on the surface.

Transformation from the geodetic coordinates $[\phi, \lambda, h]$ to the geocentric Cartesian coordinates $[X, Y, Z]$ can be performed with [2]:

$$\begin{bmatrix} X \\ Y \\ Z \end{bmatrix} = \begin{bmatrix} (N + h)\cos\phi\cos\lambda \\ (N + h)\cos\phi\sin\lambda \\ (N(1 - e^2) + h)\sin\phi \end{bmatrix} \tag{2.3}$$

where N is the *radius of curvature in the prime vertical*, which can be estimated with [2]:

$$N = \frac{a}{\sqrt{1 - e^2 \sin^2\phi}} \tag{2.4}$$

Transformation from the geocentric Cartesian coordinates $[X, Y, Z]$ to the geodetic coordinates $[\phi, \lambda, h]$ is a bit more complicated and usually an iterative loop is used to calculate the geodetic latitude ϕ and height h. By choosing an initial value of N as the semimajor axis a of the reference ellipsoid, the geodetic coordinates [ϕ, λ, h] are calculated using the following formula [2]:

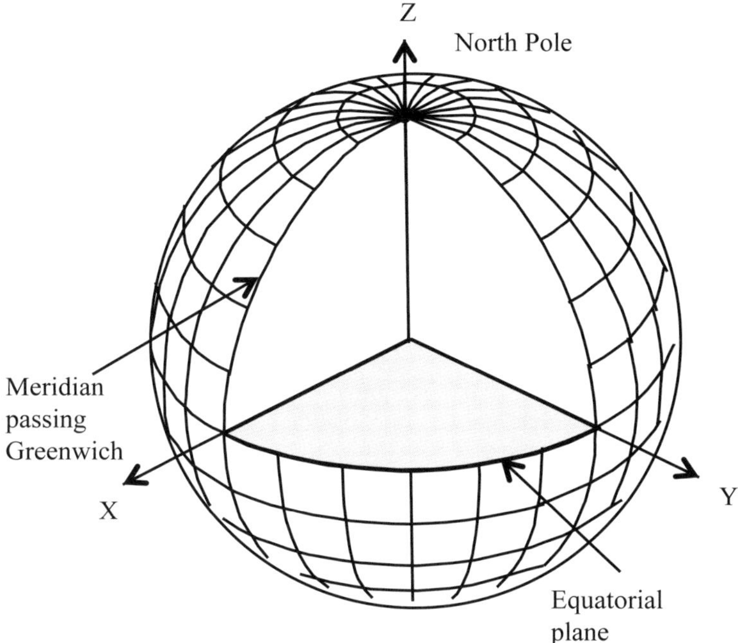

Figure 2.2 The definition of the ECEF coordinate system.

$$
\begin{bmatrix} \phi \\ \lambda \\ h \end{bmatrix} =
\begin{bmatrix}
\arctan \dfrac{Z}{\left(1 - e^2 \dfrac{N}{N+h}\right)\sqrt{X^2 + Y^2}} \\[4ex]
\arctan \dfrac{Y}{X} \\[4ex]
\dfrac{\sqrt{X^2 + Y^2}}{\cos \phi} - N
\end{bmatrix}
\tag{2.5}
$$

The results from (2.5) are used to recompute N with (2.4) and these steps are repeated until the coordinates converge to the desired precision.

2.1.3 The Local Geodetic Coordinate System

In addition to the global coordinate systems, local coordinate systems with the coordinate axes pointing to north, east, and up are commonly adopted in navigation applications, especially for pedestrian and indoor navigation. A local geodetic coordinate system is a coordinate system that has its origin at a point on the surface of the Earth. These are sometimes also known as topocentric coordinate systems, as opposed to geocentric coordinate systems, which have their origins at the Earth's center. In a local geodetic coordinate system, the z-axis is normal to the ellipsoid at the origin and pointed away from the Earth's center. The x-axis is directed toward geodetic north, and the y-axis is pointed to geodetic east, as shown in Figure 2.3 [3].

The coordinate transformation from the topocentric local geodetic coordinates $[x, y, z]$ to the ECEF coordinates $[X, Y, Z]$ can be performed with [4]:

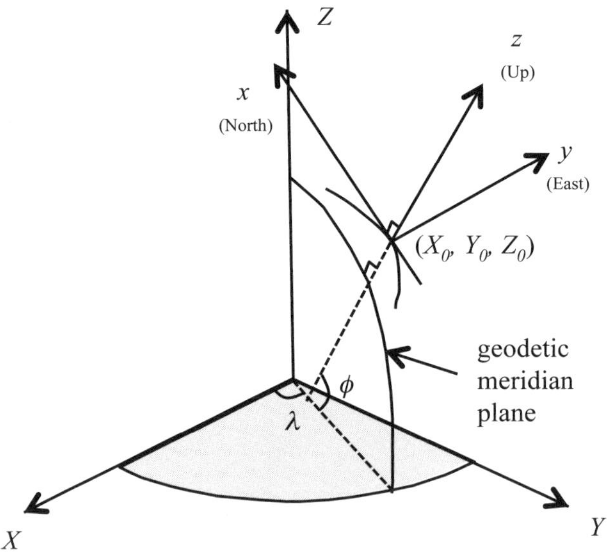

Figure 2.3 The local geodetic coordinate system.

$$\begin{bmatrix} X \\ Y \\ Z \end{bmatrix} = \begin{bmatrix} X_0 \\ Y_0 \\ Z_0 \end{bmatrix} + \mathbf{R} \begin{bmatrix} x \\ y \\ z \end{bmatrix} \quad\quad (2.6)$$

where:

$$\mathbf{R} = \begin{bmatrix} -\sin\phi_0\cos\lambda_0 & -\sin\lambda_0 & \cos\phi_0\cos\lambda_0 \\ -\sin\phi_0\sin\lambda_0 & \cos\lambda_0 & \cos\phi_0\sin\lambda_0 \\ \cos\phi_0 & 0 & \sin\phi_0 \end{bmatrix}$$

and X_0, Y_0, and Z_0 are the coordinates of the origin of the local geodetic coordinate system specified in the ECEF coordinates, while ϕ_0 and λ_0 are the latitude and longitude of the same point, respectively, in the geodetic coordinate system, as shown in Figure 2.3. The inverse transformation can be obtained easily by:

$$\begin{bmatrix} x \\ y \\ x \end{bmatrix} = \mathbf{R}^T \begin{bmatrix} X - X_0 \\ Y - Y_0 \\ Z - Z_0 \end{bmatrix} \quad\quad (2.7)$$

2.1.4 The Height System

The height system is a one-dimensional coordinate system that defines the metric distance of a point from a specified reference surface along a well-defined path (e.g., a straight line or a plumb line of the Earth's gravity field). Different reference surfaces and paths define different height systems.

The ellipsoidal height system is a purely geometrical height system. Its reference surface is an ellipsoid, as described in Section 2.1.2 and shown in Figure 2.4. The ellipsoidal height h is the metric distance measured from the surface of the ellipsoid to the corresponding point P along the normal to the ellipsoid, which is a straight line.

People do not typically use the ellipsoid height to describe elevations or altitudes but rather use the orthometric height H, as shown in Figure 2.4. For example, the elevation values shown on maps and in geospatial databases are typically taken from orthometric height measurements. Therefore, in terms of geospatial computing, it is important to understand how these values are derived.

The orthometric height represents a distance from a reference surface of an orthometric height system, known as a geoid. A geoid describes a surface of constant geopotential $W = W_0$, also known as an equipotential surface.[8] A *geoid* is designed to correspond to the global mean sea level.[9] Due to the complexity of the Earth's gravity field, however, the Earth's oceans and seas do not form a perfect ellipsoidal surface. Whereas the geometry of a reference ellipsoid is parameterized with only two parameters, the shape of the geoid may incorporate any number of parameters, depending on the desired accuracy, in order to achieve a best fit with the global mean sea level. For example, the Earth Gravitational Model 2008 (EGM2008), developed by the U.S. DoD, is a spherical harmonic model containing millions of numerical coefficients to describe the shape of the geoid.

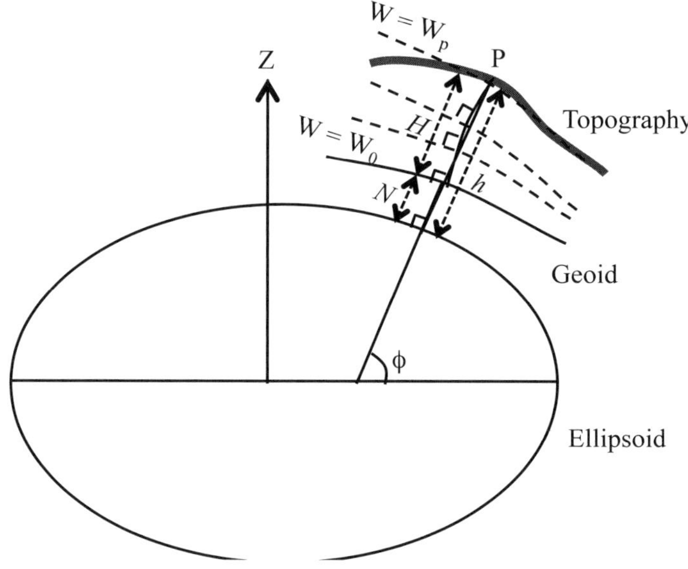

Figure 2.4 Geoid height (N), orthometric height (H), and ellipsoidal height (h).

The distance represented by the orthometric height is computed along a reference path, which unlike the ellipsoidal height, is not a straight line. In the orthometric height system, the reference path is the local plumb line of the Earth's gravity, as shown in Figure 2.4.

Orthometric heights are typically determined by spirit leveling, a technique that measures the height difference between two benchmarks. Using this

---

[8] Geopotential is the gravitational potential caused by the Earth's gravity field.

[9] Over land areas, the geoid corresponds to the surface of the sea water would theoretically form, if the continents and islands were crisscrossed with a set of tunnels and canals, allowing the sea water to extend to all regions. Thus, it is a theoretical construct, but based on physical measurements.

technique, orthometric heights can be extended over thousands of kilometers from a reference point (tidal benchmarks are typically used for such purposes).

Although the orthometric heights have been used in maps and geospatial databases, they cannot be used directly in coordinate transformations or geometry computations, such as calculating a distance in three-dimensional space between two points. The orthometric heights need to be converted to ellipsoid heights, using (2.8), before carrying out such computations:

$$h = H + N \qquad (2.8)$$

In (2.8), h and H are the ellipsoid height and orthometric height relative to a specific reference ellipsoid and geoid, respectively. N, known as the geoid height, is the distance between the reference ellipsoid and geoid, measured along the line normal to the ellipsoid. The curvature of the curved-line distance H is ignored in this equation. This approximation is acceptable for mobile positioning because the differences from the true values are well below the positioning accuracies obtainable in mobile devices.

For the reference ellipsoid and geoid defined in the WGS-84 coordinate system, the geoid height N ranges from -106.99m off the southern coast of India ($4.75°N$, $78.75°E$) to 85.39m in the Pacific Ocean north of New Guinea ($8.25°N$, $147.25°E$) [1].

2.2 POSITIONING OBSERVABLES

There are many observables that can be used for positioning mobile devices, including those based on radio signals and those that can be measured by various sensors. A sensor is a component that is capable of measuring a physical quantity by converting it into a quantifiable signal. For example, a digital camera converts photons into voltage levels, which are then converted into digital values and recorded into images. These images then become available as observables for various purposes, including positioning. Similarly, various properties of a radio frequency (RF) signal can be measured by a radio receiver and recorded as observables. One of the important types of sensors used in navigation and positioning is the inertial sensor, which includes accelerometers and gyroscopes. They are referred to as such because they measure linear and angular motion, respectively, with respect to an inertial reference frame (e.g., ECI coordinate system). Finally, digital compass is also an important sensor built-in mobile device such as a smartphone. Digital compass senses the Earth's magnetic field in order to determine the sensor orientations.

Although many of the RF signals and sensors currently used for mobile positioning were not originally designed for such purposes, due to their availability, they have been co-opted for positioning purposes [5, 6]. Thus, the built-in sensors and radio receivers in mobile devices are the components responsible for generating the positioning observables. These observables include ranges, ranging differences, speeds, accelerations, angles, angle rates, signal strengths, cell-IDs and images or images features, as listed in Table 2.1.

Table 2.1

List of Mobile Positioning Observables

Observables	*Sensors or Networks*
Range	GNSS receiver, cellular networks
Ranging difference	GNSS receiver, cellular networks
Traveled distance	Accelerometer, camera
Speed	GNSS receiver, accelerometer, camera
Acceleration	Accelerometer
Angles/azimuth	Digital compass, cellular network
Angle rates	Gyroscope
Signal strength	WLAN, Bluetooth, RFID, cellular network
Cell-ID	MAC address, base stations in cellular networks
Image/image features	Camera

In general, these observables are a function of a mobile device's coordinates as follows:

$$l = f(\mathbf{x}) + n \tag{2.9}$$

where l is the positioning observable, $\mathbf{x}$ is a vector of coordinates for one or multiple points, and n is the observable noise. For simplicity, we will ignore observable noise in this chapter, but techniques to handle this noise will be covered later in Chapters 3–5.

2.2.1 Range

A *range* is a measurement of the distance between two points. Thus, its importance for positioning should be obvious. Range observables can either be derived directly from sensors/receivers or indirectly via a model that is a function of the direct observables, for example, the path loss model of a radio channel.

In most cases, there is at least one intermediate step before obtaining a range observable. For example, the most common observable that is converted to a range observable is the time of arrival (ToA) observable. It is the signal (e.g., RF signal) traveling time from a transmitter to a receiver as follows:

$$\text{ToA} = t_r - t_t \tag{2.10}$$

where t_r is the epoch of signal reception and t_t is the epoch of signal transmission. For RF signals, the range observable d can be estimated with:

$$\begin{aligned} d &= \|\mathbf{x}_r - \mathbf{x}_t\| \\ &= \sqrt{(x_r - x_t)^2 + (y_r - y_t)^2 + (z_r - z_t)^2} \\ &= \text{ToA} \cdot c \end{aligned} \tag{2.11}$$

where c is the speed of light, (x_r, y_r, z_r) are the coordinates of the receiver, and (x_t, y_t, z_t) are the coordinates of the transmitter.

The most common range observable is the GNSS pseudorange, which is the distance between the GNSS satellite antenna and the receiver antenna (plus the offsets of the satellite and receiver clocks). Details about the GNSS pseudorange will be discussed in Chapter 3. ToA observables are also available from cellular networks, especially from code division multiple access (CDMA) networks, in which the clocks of all base stations are synchronized to an accuracy of a few microseconds [7, 8].

There are other indirect range observables, which are derived via a model that is the function of the measured observables. For example, the transmitter-receiver distance is a function of the received signal strength. In most cases, however, it is a very challenging task to model a radio channel. One issue is the complexities of the multipath components, which have complex temporal, spatial, and directional fading. In indoor environments, the transmitter-receiver distance can be approximately modeled with the following channel model (e.g., for WLAN signals [9]):

$$\text{RSS}(d) = \text{RSS}(d_0) - 10 \cdot \beta \cdot \log\left(\frac{d}{d_0}\right) - F \tag{2.12}$$

with

$$F = \begin{cases} f \cdot n & n < n_{\max} \\ f \cdot n_{\max} & n \geq n_{\max} \end{cases} \tag{2.13}$$

where d is the distance between the signal transmitter and the receiver, d_0, is the reference distance at which the reference signal strength $RSS(d_0)$ is known, $RSS(d)$ is the signal strength received at a distance d, and β is the distance-RSS gradient of the environment. F is a function representing the shadowing effects, shown in (2.13), where f is the wall attenuation factor, n is the number of walls between the signal transmitter and the receiver, and n_{max} is the maximum number of walls that can be applied to the model. Figure 2.5 shows an example of the signal attenuation patterns after penetrating different numbers of walls.

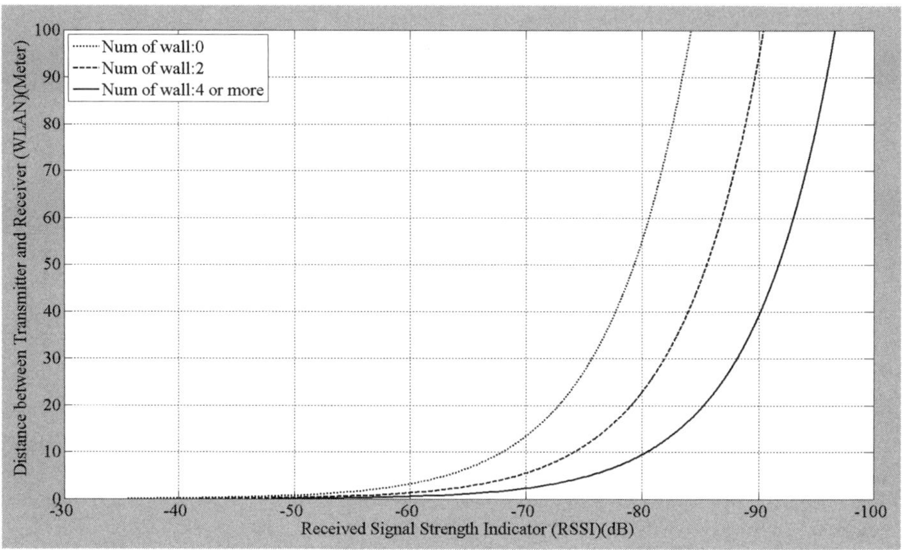

Figure 2.5 WLAN signal attenuations after penetrating walls.

2.2.2 Range Difference

A time difference of arrival (TDoA) observable is derived from the difference of two ToA observables. The two ToA observables are measured by a single receiver, which simultaneously measures the signals from two different transmitters (e.g., base stations). TDoA is related to the receiver and transmitter positions as follows:

$$
\begin{aligned}
\mathrm{TDoA} &= \mathrm{ToA}_2 - \mathrm{ToA}_1 \\
&= \frac{1}{c}(d_2 - d_1) \\
&= \frac{1}{c}\left(\sqrt{(x_r - x_2)^2 + (y_r - y_2)^2 + (z_r - z_2)^2} - \sqrt{(x_r - x_1)^2 + (y_r - y_1)^2 + (z_r - z_1)^2} \right)
\end{aligned}
\tag{2.14}
$$

where d_1 and d_2 are the distances to the first and second base stations, respectively, (x_r, y_r, z_r) are the coordinates of the receiver, (x_1, y_1, z_1) are the coordinates of the first base station, and (x_2, y_2, z_2) are those of the second base station. The major advantage of the TDoA observable is that taking the difference of two ToA observables eliminates the clock error of the receiver because the receiver clock errors are the same in both ToA observables. The two transmitters, however, must be synchronized or have a known clock offset in order for the TDoA observable to be useful.

2.2.3 Acceleration, Speed, and Traveled Distances

One type of sensor that almost all modern mobile devices have is the accelerometer. Accelerometers are typically integrated into a module containing three sensors aligned in orthogonal directions. In this way, they can provide the acceleration in three axes of the body frame of the mobile device, as shown in Figure 2.6. Measurements of the acceleration can be used for positioning and motion recognition.

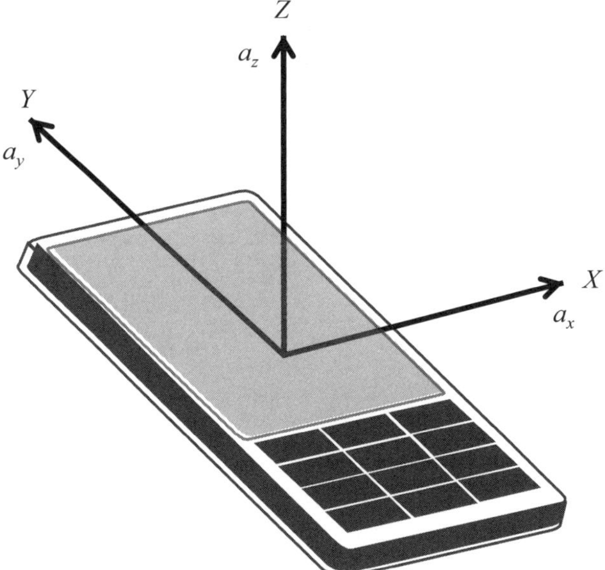

Figure 2.6 Three-axis accelerations in the body frame of a smartphone.

Theoretically, speed and traveled distance along the direction of acceleration can be derived with the following relationships:

$$v(t) = v(t_0) + \int_{t_0}^{t} a(t)dt \tag{2.15}$$

$$r(t) = r(t_0) + \int_{t_0}^{t} v(t)dt \tag{2.16}$$

where t_0 is the timestamp of the initial epoch, t is that of the current epoch, $a(t)$ is the acceleration measured by the accelerometer, $v(t_0)$ is the initial velocity, and $r(t_0)$ is the initial traveled distances that can be taken as zero in most cases. The traveled distance $r(t)$ is an observable that can be used for positioning because it is a function of the coordinates of the points along the traveling trajectory. The problem with this approach, however, is that with current accelerometer technology, there is a large noise component in the acceleration signal. Thus, large errors are introduced by the double integration resulting from (2.15) and (2.16). As a result, this is not a common approach in mobile positioning.

Instead of performing the double integration described above, the accelerometer observables are commonly adopted for positioning using other approaches: pedestrian dead reckoning and motion recognition. Both approaches avoid the problem of error accumulation exhibited by the double integration.

The pedestrian dead reckoning approach is based on the principle that, during a walking/running process, the signal of total acceleration

$$a = \sqrt{a_x^2 + a_y^2 + a_z^2} \tag{2.17}$$

elicits a waveform with a cyclic pattern, as shown in Figure 2.7. Each cycle of the waveform represents a gait cycle of the pedestrian [10, 11].

By detecting the cyclic pattern of the waveform, the number of steps walked can be counted. The traveled distance d can be estimated with:

$$d = n \cdot SL \tag{2.18}$$

or

$$d = \int_0^{\Delta t} SF(t) \cdot SL(t)dt \tag{2.19}$$

In (2.18), n is the number of walking steps and SL is the step length, which in this case is assumed to be constant. Equation (2.19) is a generalization of (2.18), in which the step length can vary with time. Here Δt is the walking duration, $SL(t)$ is the step length at time t, and $SF(t)$ is the step frequency at time t.

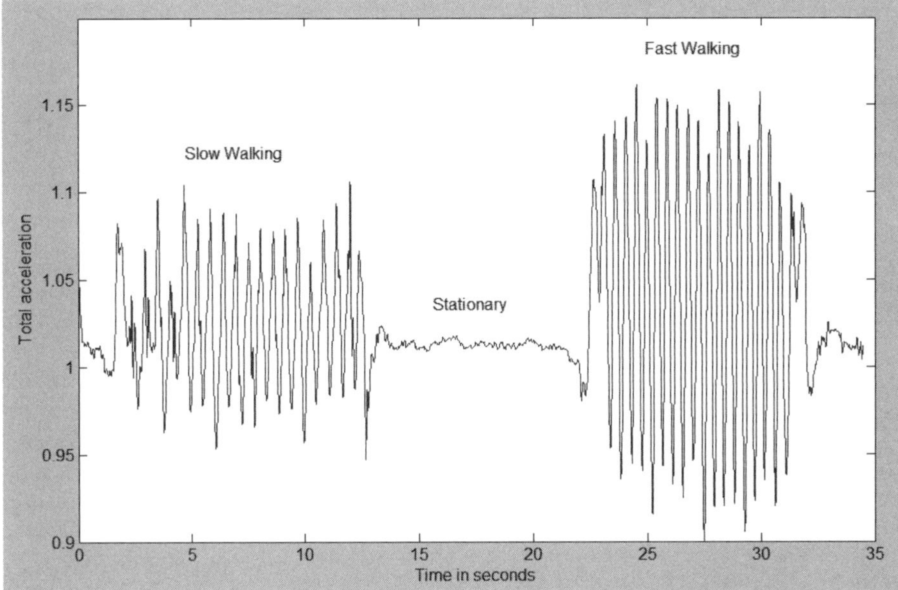

Figure 2.7 The cyclic pattern in the waveform of total acceleration during a walking process.

If the step length, step frequency, and the walking direction are all known as a function of time, the walking trajectory can be determined. Typically the step length is modeled based on training data from an individual or a group. The step frequency can be determined in real time based on the cyclic patterns measured by the accelerometer, and the walking direction can be measured with angle sensors, such as a digital compass and gyroscope. Details on this topic will be addressed in Section 2.3.3 and further in Chapter 5.

Acceleration is also one of the most common signals adapted for detecting different modes related to human locomotion, such as slow walking, fast walking, running, static, making a U-turn, and walking downstairs or upstairs [12]. These locomotion patterns can be adopted in the positioning algorithm to constrain the positioning solutions. For example, the user position should not change during the duration that the sensors have detected a "static" mode. Different walking speeds can be applied to the positioning algorithm for the durations of slow walking, fast walking, and running. Details on the topic of motion recognition will be given in Chapter 7.

2.2.4 Angles and Angle Rates

A digital compass is another common built-in sensor in mobile devices. It is used for determining the heading or azimuth of the device. The azimuth defines a straight line in the local horizontal plane (the *xy*-plane of the local geodetic coordinate system; see Figure 2.3) at a given angle from geodetic north (i.e., true north). Thus, azimuth is a function of the horizontal coordinates of two points as follows:

$$\alpha_g = \arctan\left(\frac{y_2 - y_1}{x_2 - x_1}\right) \tag{2.20}$$

where (x_1, y_1) are the horizontal coordinates in the local geodetic coordinate system of the starting point of the baseline, while (x_2, y_2) are those of the target point of the baseline.

A digital compass operates based on the principle that the horizontal component of the Earth's magnetic field always points toward the magnetic north in the local horizontal plane. The intensity of the Earth's magnetic field ranges from 0.5 to 0.6 gauss. The intensity of the horizontal component is about 0.2 to 0.3 gauss with higher intensity at the equator and lower intensity at higher latitudes. Azimuth to the magnetic north α_m can be estimated with:

$$\alpha_m = \arctan\left(\frac{H_y}{H_x}\right) \tag{2.21}$$

where H_x is the projection of the magnetic field's horizontal component in the forward direction, while H_y is its projection in the direction perpendicular to the forward direction.

The Earth's magnetic north is not identical to geodetic north. The difference between magnetic north and geodetic north is called magnetic declination. Its value varies depending on the location on the Earth, but it can be modeled or stored in a table/database.[10] The output of a digital compass is the azimuth relative to magnetic north. Thus, it has to be corrected using the magnetic declination, in order to obtain the azimuth relative to geodetic north, as follows:

$$\alpha_g = \delta + \alpha_m \tag{2.22}$$

where α_g is the azimuth relative to geodetic north, α_m is the azimuth relative to magnetic north, and δ is the magnetic declination at the current location.

One complication of using the magnetic field for positioning is that it can exhibit anomalies due to ferrous materials in the environment, and many of these

---

[10] Since the location of magnetic north changes over time, these models must be updated. For example, in aviation applications charts and databases of magnetic declination are updated twice per year. For most applications, however, an update every five to ten years would be sufficient.

are not accounted for in magnetic declination charts or databases. Especially in indoor environments, objects such as elevators, metal furniture, and even electronic equipment can cause perturbations to the magnetic field and corrupt the azimuth or heading observables.

In addition to the digital compass and accelerometer, a gyroscope is another sensor that is commonly available in modern mobile devices. Unlike the digital compass that measures magnetic azimuth, the gyroscope measures angular rate of rotation. As in the case of accelerometers, gyroscopes are typically integrated into a single module in three orthogonal directions, in order to measure rotations about three different axes. The heading or azimuth can be estimated with the angular rate output from a gyroscope as follows:

$$\alpha(t) = \alpha(t_0) + \int_{t_0}^{t} \Delta\alpha(t)\,dt \qquad (2.23)$$

where $\Delta\alpha(t)$ is the angular rate output from a gyroscope.

In addition to being measured by a digital compass or by gyroscopes, the azimuth can also be determined by measuring the direction of signal arrival in cellular networks. The direction of arrival (DoA) observable can be measured either using a directional antenna or by an array of antennae.

2.2.5 Images

High-quality cameras are nowadays widely available in mobile devices and can capture images of users' surroundings. By processing two consecutive images taken while walking, the heading change and traveled distance between these two images can be estimated [13]. Therefore, the camera functions in this case as a visual gyroscope (i.e., measuring heading change) and an odometer (i.e., measuring traveled distance). In tests using this technique, the performance of the visual gyroscope is about ± 3 degrees, while that for the visual odometer is about 2–3% of the traveled distance [14]. Although the performance of a visual gyroscope and visual odometer depends somewhat on the lighting conditions and structure of the surroundings, this technique has been found to be beneficial for positioning in many environments. Furthermore, the visual gyroscope does not have the drift problem that is exhibited with traditional gyroscopes in mobile devices. The lighting situation of the ambient environment, however, is one factor that can limit its applicability, as good lighting is essential for robustly obtaining an estimate of the heading change and travelled distance. Details of this topic will be addressed in Section 2.3.3 and additionally in Chapter 5.

2.2.6 Proximity

Proximity is a state of being near an entity. It is a qualitative measure, in that the definition of "near" is usually ambiguous. Proximity-based positioning is a technique of sensing when one is in the "surrounding" of an entity that has been associated with a location. The "surrounding" can be, for example, a signal coverage area. One must be able to identify the sensed entity using an appropriate identification scheme. Thus, the identity within such an identification scheme serves as a positioning observable. The most commonly used identification schemes for proximity-based positioning include the following:

- Cell identities (Cell-IDs or CIDs) of the base stations in cellular networks;
- Media access control (MAC) addresses of the access points of short-range RF technologies, such as WLAN, Bluetooth, and RFID.

Cell-ID is the simplest proximity-based positioning method, where Cell-ID is used to locate a mobile device within the radio coverage area of a RF transmitter in cellular networks. When a mobile device makes a connection to a base station, it is recognized by the network management system, which knows the location of the base station. Therefore, it is located automatically "within the radio coverage area" of the connected base station.

Positioning using MAC addresses is similar to the CID approach except that the location of the associated access point is available, in many cases, directly from the receiver (e.g., the smartphone). Therefore, the mobile device can locate itself using the MAC address of the connected access point without any further information from the network (although network-based approaches are also commonly used).

2.2.7 Fingerprints

A fingerprint is a set of information that represents the signal pattern of specific RF signals at a specific location in a physical environment. Based on the assumption that different locations have different patterns, a mobile device can be located by finding the best match of the observed signal pattern to those stored in a fingerprint database. An RSS radio map (or database) is the most common way to store fingerprints. Fingerprinting with RSS radio maps has been widely used for both indoor and outdoor positioning [9]. Radio maps can be generated outdoors for cellular networks and indoors for WLAN, Bluetooth, and RFID networks.

In addition to radio maps, images or features in images (lines, corners, et al.) can be used as fingerprints for positioning in mobile devices as well, especially for indoor environments.

There are many algorithms for "finding the best match" between the observed signal pattern and the stored fingerprints. More details on this topic will be discussed in Chapter 4.

2.3 POSITIONING METHODS

Positioning methods can be divided into two high-level categories: absolute positioning and relative positioning. In absolute positioning, the target's position is expressed directly with respect to a coordinate system, such as geocentric coordinates (x, y, z). In relative positioning, the target's position is expressed relative to another known position. Positioning methods for mobile devices can be divided into three categories: GNSS positioning, positioning based on the RF signals of wireless communications networks, and hybrid positioning. Hybrid positioning is defined as any positioning method combining two or more separate positioning methods. Hybrid positioning can include relative positioning methods, since their combination with, for example, an initial position, can result in an absolute position.

2.3.1 GNSS Positioning

GNSS is nowadays the most common positioning method, especially for outdoor environments where open views to sky are available. Its applications range from professional ones, such as surveying and monitoring of crustal deformations, to car navigation and other mobile location-based services. The positioning accuracy of GNSS ranges from a few millimeters (using high-end geodetic class receivers) to a few meters [using low-cost receivers (e.g., the GNSS receivers in smartphones)].

In addition to standalone GNSS positioning, there are several other GNSS positioning methods that utilize additional components to enhance the performance in various ways, including assisted-GNSS (A-GNSS) positioning, differential GNSS (DGNSS or DGPS) positioning, and positioning based on a satellite-based augmentation system (SBAS).

2.3.1.1 The GNSSs

At the time of this writing, there are two operational GNSSs: the U.S. GPS and the Russian GLONASS. In addition to these two operational systems, two other systems are now under development. One is the European Galileo system, while the other is the Chinese system Compass. All GNSSs require the use of multiple Earth-orbiting satellites. The set of satellites belonging to a specific GNSS provider is collectively known as a *constellation*.

GPS

The nominal GPS constellation consists of 24 satellites orbiting in six planes that have an inclination angle of 55 degrees. The orientation of the planes is such that they are equally spaced around the equator at a 60-degree separation. The GPS orbits are nearly circular with a radius of approximately 26,600 kilometers from the Earth's center of mass. The orbital period is half of a sidereal day, or 11 hours 58 minutes [15].

The legacy GPS satellites transmit three signals, called L1 C/A, L1P(Y), and L2P(Y), at two carrier frequencies: L1 and L2. The center frequency of L1 is 1.57542 GHz, while that of L2 is 1.2276 GHz. The L1 C/A signal is freely accessible for public services, while the L1P(Y) and L2P(Y) signals are encrypted signals for military use. GPS has been fully operational since 1995. The system is now undergoing modernization, including the addition of a new carrier frequency, L5, and in total five new signals, including: L5, L1M, L2M, L1C, and L2C. The center frequency of L5 is 1.17645 GHz. GPS is a CDMA system, where the signals from different satellites are identified by different pseudo random noise (PRN) codes. These codes have the appearance of being a binary random sequence, but in fact they are fully deterministic.

GLONASS

The nominal GLONASS constellation consists of 24 satellites orbiting in three planes that have an inclination angle of 64.8 degrees. GLONASS satellites transmit two signals, an open standard precision (SP) signal and a restricted high precision (HP) signal, in two carrier frequencies L1 and L2, respectively [12]. Unlike the GPS system, GLONASS is a frequency division multiple access (FDMA) system, where signals transmitted from different satellites are discriminated by different frequencies (instead of by different PRN codes as in the case of GPS). The frequencies assigned to different satellites can be calculated in gigahertz according to [16]:

$$f_{L1}^{k} = 1.602 + 0.0005625 \cdot k$$
$$f_{L2}^{k} = 1.246 + 0.0004735 \cdot k$$

$$(2.24)$$

where $k = \{-7, -6, \ldots 5, 6\}$ is the channel number. Although GLONASS was fully operational beginning in 2011, a modernization process is also ongoing for the system. One of the most significant trends in this modernization process is that GLONASS is shifting to a code division multiple access (CDMA) system so that it will be more compatible with other GNSSs. A CDMA signal L3OC was added in 2011 to a new frequency band L3, which ranges from 1.197648 GHz to 1.212255 GHz [16].

Galileo

The nominal constellation of the European GNSS Galileo will consist of 30 satellites orbiting in three planes. These planes have an inclination angle of 56 degrees, which is similar to that of GPS. The Galileo satellites will transmit various signals in four frequency bands, namely E1, E5a, E5b, and E6, to support the open service (OS), safety-of-life service (SoL), commercial service (CS), public regulated service (PRS), and search and rescue (SAR) service, respectively. The center frequencies are 1.57542 GHz for E1, 1.19175 GHz for E5 and 1.27875 GHz for E6 (which despite their names, are located in the L-band).

Compass

Compass is a GNSS being developed in China. The implementation of the Compass system consists of three phases [17]:

- Phase I: BeiDou Navigation Satellite Demonstration System, which was established in 2000.
- Phase II: Regional BeiDou Navigation Satellite System, which has provided service for China and its surrounding areas since 2012.
- Phase III: Global BeiDou Navigation Satellite System, which will provide global service by 2020.

The nominal constellation of the regional BeiDou Navigation Satellite System (Phase II) consists of 14 satellites, of which five are geostationary Earth orbit (GEO) satellites, four are medium Earth orbit (MEO) satellites, and five are inclined geosynchronous orbit (IGSO) satellites. These satellites will transmit the B1 signal at a center frequency of 1.561098 GHz, which is slightly different from that of the GPS L1 and Galileo E1 signals.

According to available knowledge at the time of this writing, the global constellation of the Compass system (Phase III) will include 35 satellites, of which 27 are MEO satellites, five are GEO satellites, and three are IGSO satellites [18]. As mentioned above, the system will be operational by 2020. Therefore, it is possible that the final constellation may differ slightly from the current definition.

2.3.1.2 Standalone GNSS Positioning

GNSS positioning is based on trilateration using ranging observables (ρ_i) from the receiver to multiple GNSS satellites, as shown in Figure 2.8.

The positions of the satellites can be calculated with the satellite ephemerides obtained from the navigation messages. The three-dimensional position of the receiver (x, y, z) can be estimated, theoretically, with three ranging observables. However, the clock in the receiver is not synchronized with GNSS time; therefore, it has to be treated as an additional parameter in the position estimation process. Therefore, ranging observables to four satellites are required in order to estimate the parameter tuple (x, y, z, Δt), where Δt is the clock offset of the receiver from the GNSS time. More details on this topic will be addressed in Chapter 3.

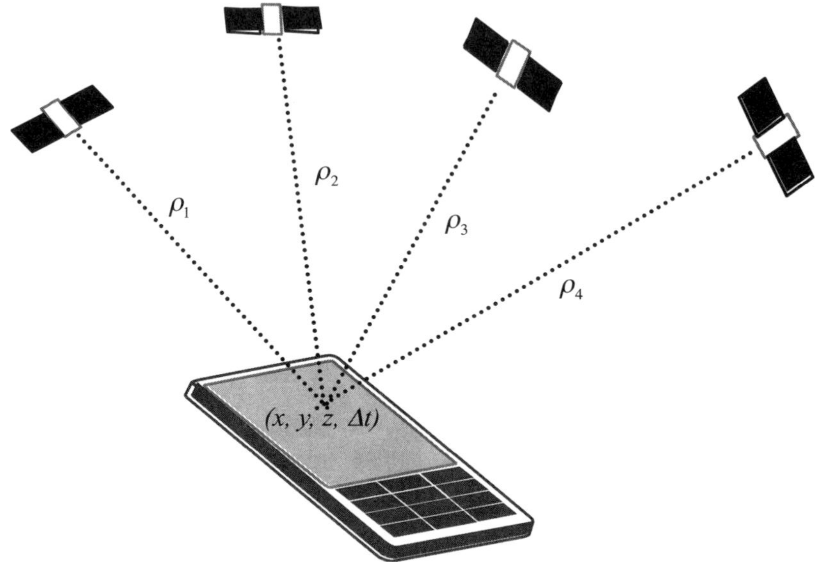

Figure 2.8 Concept of GNSS trilateration.

2.3.1.3 Assisted GNSS Positioning

As described above, satellite positions (calculated with the ephemerides) are required for positioning. The ephemerides are broadcasted at a rate of 50 bits per second (bps) in the legacy GPS signals. For the best case (when there are no bit errors), the reception time of a complete navigation dataset, which includes the satellite ephemeris, system time information, and satellite health status, will be 18 seconds. In practice, GNSS signals are attenuated significantly in challenging environments, such as so-called urban canyons, and bit errors occur frequently under such conditions. As a result, it can take minutes, or even tens of minutes, to receive a complete navigation dataset. This will prolong the Time-To-First-Fix (TTFF), which is the time from switching on the receiver to obtaining the first position.

A-GNSS is designed to overcome this problem by delivering the navigation dataset to mobile users via a data communication link from cellular networks or another data connection. In this case, mobile devices can start offering positions without waiting a long period for the reception of the navigation dataset from satellites. In addition to the benefits of speeding up the TTFF process, A-GNSS can also enhance the tracking sensitivity of the receiver by removing the navigation data bits from the received signals and extending the duration of coherent signal integration. More details of this topic will be addressed in Chapter 3.

2.3.1.4 Differential GNSS Positioning

When GNSS signals propagate through the atmosphere to the ground, they suffer from refractions in the ionosphere and troposphere. In addition, the signals may reflect off objects in the surroundings of the receiver and thus arrive to the receiver along different paths, producing so-called multipath effects. Additionally, the pseudorange observable is affected by errors of the satellite clock, uncertainty of the satellite orbits, and the hardware bias of the receiver. Some of these errors, such as ionospheric delays, tropospheric delays, and satellite orbit and clock errors are highly correlated in space and time, meaning that measurements taken at nearby locations and close to each other in time will have similar errors. The differential GNSS (DGNSS) solution takes advantage of this phenomenon, utilizing a reference receiver at a known location and a radio broadcast link. The reference receiver determines pseudorange corrections, consisting primarily of the spatially and temporally correlated errors. These corrections are then broadcast to other end-user receivers within the effective range of the reference base station (e.g., 150 kilometers). The end-user receivers apply these corrections to the observed pseudoranges, remove many of the errors described above and allow improving the positioning accuracy. Further explanation of this topic will be presented in Chapter 3.

2.3.1.5 SBAS

The original GPS signals do not contain any information to facilitate the estimation of the error bounds of the estimated positions, which is important for SoL applications such as precision approach of aircraft. This was the motivation for developing SBASs. Similar to the differential GNSS, a SBAS provides end-users with ionospheric corrections, satellite clock corrections, and satellite orbit corrections. Unlike the DGNSS, the SBAS signals are transmitted by geostationary (GEO) satellites. In addition, SBAS delivers standardized integrity information, which indicates the "level of trust" (such as error bounds) that can be applied to the GNSS signals. The integrity information enables the end-users to estimate the protection levels (PLs) of the calculated positions.[11]

SBAS is essentially a regional system; a number of systems have been implemented in different regions of the world. These regional SBASs include the U.S. wide area augmentation system (WAAS), the European geostationary navigation overlay service (EGNOS), the Japanese multifunctional transport satellite (MTSAT) based satellite augmentation system (MSAS), and the Indian GPS and GEO augmentation navigation (GAGAN) system. Further details of these topics will be discussed in Chapter 3.

---

[11] Protection levels are formally defined error bounds for the estimated position, used especially in aviation applications.

2.3.2 Positioning Based on RF Signals of Wireless Networks

Whereas GNSS signals were designed specifically for the purposes of positioning, many other RF signals (designed for other purposes) exist within the ambient environment and can be received by mobile devices, most notably signals from cellular networks but also signals from WLAN and Bluetooth access points. Many observables from these RF signals can be utilized for mobile positioning. These observables include the cell identity of the base stations of cellular networks, the MAC address of the access points of short-range RF networks, the received signal strength or received signal strength indicator (RSSI), the direction (or angle) of arrival, the ToA, and the TDoA. Positioning methods using these observables include radio signal coverage area, fingerprinting, trilateration, and triangulation. Assisted GNSS can also be considered as one of the positioning methods in this category for the following reasons: 1) the GNSS receiver is now a standard positioning component in most mobile devices; 2) the operation of an A-GNSS service requires support from cellular networks; and 3) the A-GNSS position method is included in the standards of cellular networks.

2.3.2.1 The Signal Coverage Area Method

The basic principle of this approach is to detect the RF transmitter to which the mobile device is currently connected. The position of the mobile user is then associated with the signal coverage area of the corresponding RF transmitter. Depending on the antenna types of the RF transmitters, the shapes of the signal coverage areas are different. The most common signal coverage areas include the following:

- A circle when the RF transmitter has an omni antenna;
- A sector of a circle when the RF transmitter has a directional antenna.

Typically, the position estimated with this approach is the location of the corresponding RF transmitter. This is called the Cell-ID approach.

Sometimes the potential position of a mobile device is further shrunk to a particular part of the signal coverage area by using additional constraints, such as the received signal strengths. This part of signal coverage area is the potential area in which the mobile device is currently located. For example, the combination of received signal strength and the signal coverage area of a RF transmitter with a directional antenna will define a potential area that can be much smaller than the original signal coverage area (e.g., the entire circular sector). The size of the potential area is dependent on the accuracy of the constraint observation. For example, the error of the received signal strength defines the minimum and maximum radii of the potential area. This is called the enhanced Cell-ID approach. The position estimated with the enhanced Cell-ID approach is the location of the centroid of the potential area, while the positioning accuracy is the maximum distance from the centroid to the boundary of the potential area.

2.3.2.2 Fingerprinting

The fingerprinting method is based on the fact that some observables, such as receiver signal strength, are spatially correlated. If the fingerprints (i.e., signal patterns) at "reference locations" are known in advance, it is possible to locate the receiver by matching the observed signal pattern with the known fingerprint at a "reference location."

The fingerprinting approach has two phases: the training phase and the positioning phase. The training phase is required to collect the fingerprints at different reference locations in the target area, while the positioning phase is to find a reference location whose fingerprint has the best match to the observed signal pattern. The best match can be defined as, for example, the shortest signal distance. Fingerprinting is a very popular positioning method for indoor environments, where the RSSI observables from WLAN, Bluetooth, or RFID networks are used as the fingerprints. Further details on this topic can be found in Chapter 4.

2.3.2.3 Trilateration

In order to constrain the user position, we can use the fact that a single range measurement defines a circle with a radius equal to the corresponding range. Two circles in a plane can intersect at most at two different points, and three circles can together intersect at only a single point, as shown in Figure 2.9. This is the basic principle of the trilateration positioning method. It is based on the intersection of three or more circles, where range measurements define the radii of the circles. In practice, there are errors in the ToA measurements from which the range measurements are derived. Therefore, the intersection of the circles is not a single point but rather a roughly triangular error space, as shown in Figure 2.9. The area of the error space depends on the sizes of the errors in the ToA measurements.

In addition to ToA measurements, TDoA measurements can be used for trilateration as well. A TDoA measurement defines a hyperbola. The position of a mobile device can be obtained by the intersection of two hyperbolae. More details on this topic are addressed in Chapter 4.

2.3.2.4 Triangulation

As mentioned in Section 2.2.4, a DoA observable (azimuth) can be used to define a straight line in the xy-plane of the local geodetic coordinate system, while two nonparallel straight lines in the xy-plane intersect to a single point, as shown in Figure 2.10. Based on this principle, a mobile device can be located by triangulation using two or more DoA measurements. Similar to the trilateration approach, an error space is introduced by the errors of the DoA measurements (ΔA in Figure 2.10).

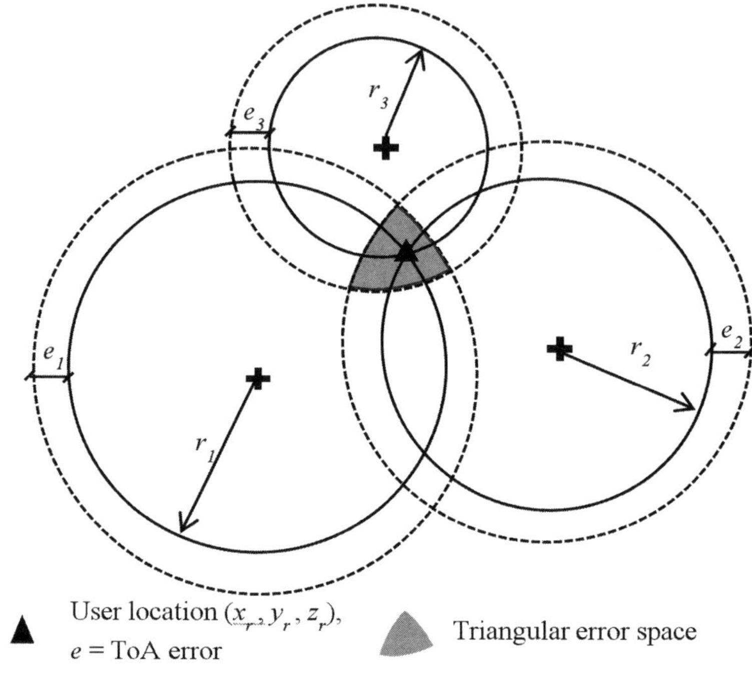

Figure 2.9 Trilateration with three ToA measurements.

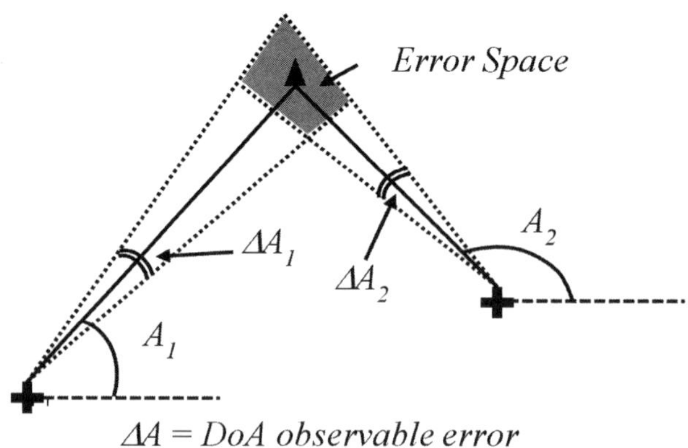

Figure 2.10 Triangulation with two DoA measurements.

2.3.3 Hybrid Positioning

As shown in Table 2.1, a mobile device has not only multiple built-in sensors but also a variety of radio interfaces for facilitating wireless data communications. Observables derived from these sensors and radio interfaces offer the opportunity to develop hybrid positioning solutions that are based on a multisensor, multinetwork approach. Hybrid positioning solutions benefit from the advantages of each sensor and RF signal. They offer better positioning accuracy and availability compared to positioning solutions based on a single technology.

Hybrid positioning solutions may include the following:

- Multiple sensors integration;
- Multiple sensors and multiple RF signals integration;
- Pedestrian dead reckoning (PDR);
- Visual positioning.

2.3.3.1 Multiple Sensor Integration

Sensor-based hybrid positioning solutions require the integration of two types of sensors:[12]

- Type 1: Sensors that can provide ranging or directional observables, which can be used to directly determine absolute positions;
- Type 2: Sensors that can provide observables for relative positioning at a high repetition rate.

The GNSS receiver is the most common type 1 sensor, providing range observables for absolute positioning, while accelerometers, gyroscopes, and odometers are common type 2 sensors, used to derive the speed, heading change, and traveled distance. Although type 2 sensors cannot be used as standalone sensors for positioning, they can propagate positions from an initial position for estimating a trajectory from a known location (i.e., relative positioning). The major benefits of integrating type 2 sensors into a hybrid positioning solution are described as follows:

- They can provide positioning solutions during a short period of outage of type 1 sensors (for example, during passing through a tunnel where GNSS positioning solutions are not available).
- They can typically offer positions at a higher rate for high-dynamic applications (e.g., driving on a highway). Most type 2 sensors have a high

---

[12] The names given to these sensor types (type 1/type 2), adopted for conciseness, are arbitrarily assigned.

data rate. Therefore, integrating them with type 1 sensors will enable positioning solutions at a higher rate.

Positioning algorithms for multiple sensor integration will be explained in detail in Chapter 5.

2.3.3.2 Multiple Sensors and Multiple RF Signals Integration

One limitation of the GNSS-based positioning solutions is that GNSS signals do not penetrate into deep indoor environments. Therefore, we need to find a replacement of this method for such environments. Pseudolites (RF transmitters that transmit a GNSS-like signal) are one alternative, while short-range RF communication technologies (e.g., WLAN) are another.

Adopting pseudolite technology for indoor environments will enable a seamless indoor/outdoor GNSS-based positioning solution because only minor modification to the receiver firmware is needed to support the tracking of both GNSS satellites and pseudolites [19]. The accuracy of pseudolite-based positioning solutions indoors, however, is low because the pseudolite signals suffer seriously from multipath effects. In addition, the installation of pseudolite systems is neither a common nor a cost-effective approach for most indoor environments. Furthermore, in most countries the use of transmitters in the RF spectrum used for GNSS is closely regulated.

Instead, WLAN is a technology that has worldwide acceptance as a short-range communication technology. The deployment of WLAN networks has covered most public areas such as airports, convention centers, museums, restaurants, hotels, hospitals, schools, university campuses, and shopping malls. Use of RSS observables of WLAN networks is a good alternative, especially when integrated with high-rate relative positioning sensors, to achieve a better positioning availability in indoor environments.

From a theoretical standpoint, there are various approaches for fusing the observables from type 1 and type 2 sensors. These approaches can be implemented easily with a *Kalman filter* [5, 6]. More details on this topic, especially the possible positioning algorithms, will be addressed in Chapter 5.

2.3.3.3 PDR

One of the major disadvantages of inertial sensors, such as gyroscopes, is their error that accumulates with time, known as drift. As a result, these sensors can be used to propagate positions only for a very short period. They need to be calibrated frequently with absolute positioning sensors (i.e., type 1) such as GNSS receivers. The effective period for propagating positions varies depending on the quality of the sensors.

One solution to avoid the drift problem of inertial sensors is the PDR solution, where one calculates the current position from the previously obtained position, using the measured speed and direction of motion. For the case of pedestrian motion, walking speed can be inferred from the acceleration pattern as described in Section 2.2.3, because it is the product of the step frequency and step length. With knowledge of the step length, step frequency, and the walking direction, the walking trajectory can be determined [10]. Details of the PDR algorithm will be addressed in Chapter 5.

2.3.3.4 Visual Positioning

Another built-in sensor in mobile devices that can be used for positioning is the camera. As described in Section 2.2.5, the positioning observable of the camera is the image or the features extracted from images. These observables have very different characteristics than that of the inertial sensors. Images do not have a drift problem and are not affected by ferrous materials existing in the environment. They are, however, affected by the lighting conditions and availability of useful image features (further described below).

Visual positioning has been widely adopted for robot navigation; however, it has been investigated only recently for pedestrian navigation in mobile devices [13, 14]. There are two approaches for visual positioning: image matching and camera motion detection. The first approach offers an absolute position (i.e., type 1 sensor), while the second approach is an aiding solution, which has to be integrated with other sensors in order to obtain an absolute position (i.e., type 2 sensor). Figure 2.11 shows the basic data processing procedure of image matching and camera motion detection.

The data processing procedure includes three major steps: image pre-processing, feature extraction, and positioning. The objective of the image pre-processing step is to remove the image noise by applying a filter to the original image (e.g., a linear filter with a Gaussian kernel).

Feature extraction is an important step in visual positioning or visual-aided positioning. Lines are the most common features that can be extracted from an image for such purposes. Line extraction typically includes two steps: edge detection with, for example, the Canny algorithm [20], and line detection using a Hough transformation [21, 22]. In addition to lines, corners are common features in images, especially those taken indoors or in urban canyons. Corners can be detected and extracted with the Harris corner detector [23]. For images that do not have straight lines and corners, the scale invariant feature transform (SIFT) algorithm can be adopted for detecting other features in the image [24]. A feature detected by the SIFT algorithm is described by a local feature vector called a SIFT key. A SIFT key is invariant to image translations, scaling, rotation, and, to some extent, illumination changes.

Image matching can be performed either by comparing the whole image or special features, which can be lines, corners, or SIFT keys. It is similar to the fingerprinting technique described in Section 2.2.7, as the location is determined by finding the best match of the query image (the one captured in real time) from a set of reference images stored in a database.

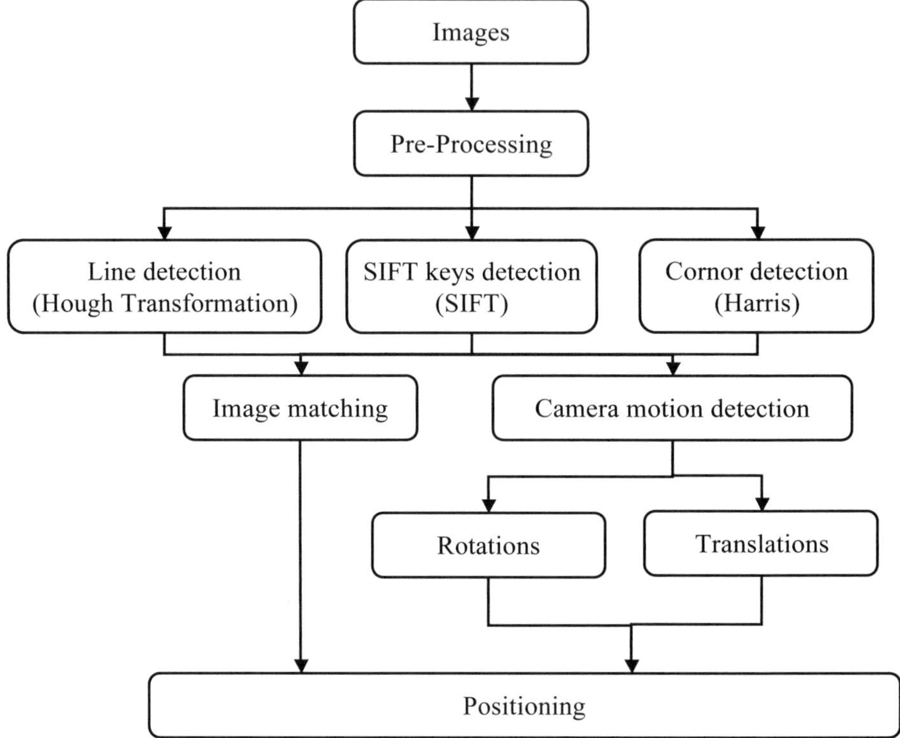

Figure 2.11 Data process diagram of visual positioning.

With regard to camera motion detection, two types of camera motion can be detected using consecutive images: rotation and translation. The camera rotation between two consecutive images can be estimated by calculating the coordinates of vanishing points (where two parallel lines appear to intersect) in these two images. The key for estimating the translation, on the other hand, is the determination of image depth, which can be calculated using the following three approaches:

- Using images from stereo cameras;
- Applying known dimensions of real-world objects;
- Tilting the camera slightly downward to the floor with a known camera height and tilt angle. This approach only works indoors with a flat floor [9].

Camera rotation is equivalent to the output of a gyroscope, while translation is equivalent to the output of an odometer. This is why implementations of these techniques are sometimes called visual-gyroscope and visual-odometer, respectively. They reveal the relative change in position between two consecutive images. Although they cannot be used for absolute positioning, they can be integrated with other sensors and/or RF signals for determining the camera position. More details with regard to visual positioning algorithms will be discussed in Chapter 5.

2.4 SUMMARY

This chapter described the fundamentals of mobile device positioning. It introduced the relevant coordinate systems, explained various observable types and observables that are available in mobile devices for positioning, and briefly described the positioning methods that utilize these observables from different built-in sensors and RF signals. All these positioning methods will be expanded upon in further detail, especially in Chapters 3–5.

References

[1]   National Imagery and Mapping Agency. World Geodetic System—Its definition and relationships with local geodetic systems. NIMA technical report TR8350.2, 2000.

[2]   Torge, W., *Geodesy,* Berlin|New York: Walter de Gruyter, 1980.

[3]   Vanicek, P., and E. Krakiwsky, *Geodesy, the Concept,* Amsterdam|NewYork|Oxford: North-Holland, 1982.

[4]   http://en.wikipedia.org/wiki/Geodetic_system

[5]   Chen, R., *Ubiquitous Positioning and Mobile Location-Based Services in Smart Phones,* Pennsylvania: IGI-Global, 2012.

[6]   Chen, R., "Introduction to smart phone positioning," In *Ubiquitous Positioning and Mobile Location-Based Services in Smart Phones,* pp. 1-31, R. Chen (ed), Pennsylvania: IGI-Global, 2012.

[7]   Diggelen, F. V., *A-GPS, Assisted GPS, GNSS and SBAS,* Norwood, MA: Artech House, 2009

[8]   Chen, R., "Assisted GNSS in smart phones," In *Ubiquitous Positioning and Mobile Location-Based Services in Smart Phones,* pp. 32-43, R. Chen (ed), Pennsylvania: IGI-Global, 2012.

[9]   Chen, R., et al., "WLAN and Bluetooth Positioning in Smart Phone," In *Ubiquitous Positioning and Mobile Location-Based Services in Smart Phones,* pp. 44-68, R. Chen (ed), Pennsylvania: IGI-Global, 2012.

[10] Chen, W., et al., "Comparison of EMG-based and Accelerometer-based Speed Estimation Methods in Pedestrian Dead Reckoning," *Journal of Navigation,* 64, 2011, pp. 265–280.

[11] Chen, R., et al., "Sensing Strides Using EMG Signal for Pedestrian Navigation," *GPS Solutions,* 15, 2011, pp. 161-170.

[12] Pei, L., et al., "Using LS-SVM Based Motion Recognition for Smartphone Indoor Wireless Positioning," *Sensors,* 12(5), 2012, pp. 6155-6175.

[13] Ruotsalainen, L., and H. Kuusniemi, "Visual Positioning in a Smartphone," In *Ubiquitous Positioning and Mobile Location-Based Services in Smart Phones*, pp. 130-158, R. Chen (ed), Pennsylvania: IGI-Global, 2012.

[14] Ruotsalainen, L., "Visual Gyroscope and Odometer for Pedestrian Indoor Navigation with a Smartphone," *Proc. ION GNSS 2012*, Nashville, Tennessee, Sept. 17-21, 2012.

[15] Kaplan, E. D., and C. J., Hegarty (eds). *Understanding GPS, Principles and Applications.* Second Edition, Norwood, MA: Artech House, 2006.

[16] http://en.wikipedia.org/wiki/GLONASS#CDMA_signals

[17] "BeiDou Navigation Satellite System Signal In Space Interface Control Document (Test Version)." China Satellite Navigation Office, 2011.

[18] http://en.wikipedia.org/wiki/Beidou_navigation_system

[19] Chen, R., et al., "Development of the EGNOS Pseudolite System," *Journal of Global Positioning Systems,* Vol. 6, No:2, 2008, pp. 119-125.

[20] Canny, J. F., "A Computational Approach to Edge Detection," *IEEE Transactions on Pattern Analysis and Machine Intelligence,* 8(6), 1986, pp. 679-698.

[21] Hough, P.V.C., "Method and Means for Recognizing Complex Patterns," *U.S. Patent No. 3069654.* Washington DC: U.S. Patent and Trademark Office, 1962.

[22] Duda, R. O., and P. E. Hart, "Use of the Hough Transformation to Detect Lines and Curves in Pictures," *Communications of the ACM, 15,* 1972, pp. 11-15.

[23] Harris, C., and M. Stephens, "A Combined Corner and Edge Detector," *Proc. the Fourth Alvey Vision Conference*, 1988, pp. 147-152.

[24] Lowe, D. G., "Object Recognition From Local-Scale Invariant Features," *Proc. the International Conference on Computer Vision*, Corfu, Greece, 1999, pp. 1150-1157.

Chapter 3

GNSS Positioning in Mobile Devices

The GNSS is now becoming the most popular choice for positioning technology, due to its high positioning accuracy, high availability, cost effectiveness, and technological maturity. As mentioned in Chapter 2, there are various positioning solutions utilizing GNSS technology that are relevant to the low-cost receiver built into mobile devices. These solutions include the standalone positioning solution, the DGNSS positioning solution, the SBAS positioning solution, and the A-GNSS positioning solution. Depending on the positioning environment, the availability of error corrections, and wireless communication connectivity, different position solutions can be applied for different situations. The DGNSS positioning solution provides a better positioning accuracy by applying the error corrections offered by a service provider. Therefore, it is a desirable solution whenever and wherever the service is available. An SBAS typically provides less accurate error corrections (compared to differential GNSS corrections); however, it is a free-of-charge service without any additional hardware required in the GNSS receiver (assuming software support for SBAS). Furthermore, SBAS includes integrity information for estimating the error bounds of the calculated positions. Therefore, it is an important solution for two types of applications: SoL applications, such as aircraft landings, and liability critical services, such as toll road pricing, car insurance, and legal applications. Although mobile devices are not involved in any SoL applications now or in the foreseeable future, liability critical services have a high potential to be important applications in future mobile devices, especially for applications such as toll road pricing and car insurance. A-GNSS solutions offer assistance data over a wireless data connection to speed up the TTFF process—the time span from the moment when the user turns on the receiver to the moment that the receiver delivers the first position. It is important to use this technology especially when positioning is carried out in urban environments, where open views to sky are often not available. This chapter will describe these positioning solutions in detail, including the positioning algorithms, estimation of the position accuracies, and integrity assessment for the case of the SBAS assisted positioning solution.

3.1 STANDALONE GNSS POSITIONING

The standalone positioning solution requires only visibility to the GNSS satellites. As long as an open view to the sky is available, the GNSS receiver can track the satellites in view and determine the user's position (assuming no anomalous conditions like a major geomagnetic storm). As was shown in Section 2.7, determining the three-dimensional coordinates and the receiver clock offset of a smartphone requires four pseudorange measurements. A pseudorange is the geometric distance between the receiver and the satellite plus the satellite and receiver clock offsets. Pseudorange is the major observable delivered by GNSS receivers. It is also the most common observable used for positioning in mobile devices.

3.1.1 Calculation of Satellite Positions and Clock Offsets

Satellite orbits and clock offsets are typically estimated by the ground control segment of a GNSS and delivered to the users via satellite ephemerides. Table 3.1 lists some of the parameters contained in an ephemeris. These parameters are used to estimate the satellite position and clock offsets, which are fundamental information for GNSS positioning.

Table 3.1

Parameters for Estimating Satellite Position and Clock Offset

Parameters	*Description*
t_{0e}	Reference time of ephemeris
$\sqrt{a}$	Square root of the semimajor axis
e	Eccentricity
i_0	Inclination angle at reference time
l_0	Longitude of the ascending node at weekly epoch
ω	Argument of perigee at reference time
M_0	Mean anomaly at reference time
di/dt	Rate of change of the inclination angle
$\dot{\Omega}$	Rate of node's right ascension
Δn	Mean motion corrections
C_{us}	Amplitude of sine correction to argument of latitude
C_{uc}	Amplitude of cosine correction to argument of latitude
C_{rs}	Amplitude of sine correction to orbital radius
C_{rc}	Amplitude of cosine correction to orbital radius
C_{is}	Amplitude of sine correction to inclination angle
C_{ic}	Amplitude of cosine correction to inclination angle
a_0	Polynomial coefficient for satellite clock correction
a_1	Polynomial coefficient for satellite clock correction
a_2	Polynomial coefficient for satellite clock correction

Table 3.2 lists the algorithm for estimating the satellite positions using the ephemeris parameters.

Table 3.2

Algorithm for Calculating Satellite Position in WGS-84 Coordinate System

Steps	*Description*
$a = (\sqrt{a})^2$	Semimajor axis
$n = \sqrt{\dfrac{\mu}{a^3}} + \Delta n$	Corrected mean motion, $\mu = 3986004.418 \cdot 10^8\, m^3 s^{-2}$
$t_k = t - t_{0e}$	Time from ephemeris epoch (GPS time)
$M_k = M_0 + n t_k$	Mean anomaly
$E_k = M_k + e \sin E_k$	Eccentric anomaly (solved iteratively for E)
$\sin v_k = \dfrac{\sqrt{1-e^2}\,\sin E_k}{1 - e\cos E_k},\quad \cos v_k = \dfrac{\cos E_k - e}{1 - e\cos E_k}$	True anomaly
$\phi_k = v_k + w$	Argument of latitude
$\delta\phi_k = C_{us}\sin(2\phi_k) + C_{uc}\cos(2\phi_k)$	Argument of latitude correction
$\delta r_k = C_{rs}\sin(2\phi_k) + C_{rc}\cos(2\phi_k)$	Radius correction
$\delta i_k = C_{is}\sin(2\phi_k) + C_{ic}\cos(2\phi_k)$	Inclination correction
$u_k = \phi_k + \delta\phi_k$	Corrected argument of latitude
$r_k = a(1 - e\cos E_k) + \delta_r$	Corrected radius
$i_k = i_0 + (di\,/\,dt)t_k + \delta i_k$	Corrected inclination
$l_k = l_0 + (\dot{\Omega} - \dot{\Omega}_e)t_k - \dot{\Omega}_e \cdot t_{0e}$	Corrected longitude of node. $\dot{\Omega}_e$ is the Earth rotation rate: $7.2921151467\text{x}10^{-5}\ rad/s$
$x_p = r_k \cos u_k$	Orbit plane x coordinate
$y_p = r_k \sin u_k$	Orbit plane y coordinate
$x_s = x_p \cos l_k - y_p \cos i_k \sin l_k$	WGS-84 X coordinate
$y_s = x_p \sin l_k + y_p \cos i_k \cos l_k$	WGS-84 Y coordinate
$z_s = y_p \sin i_k$	WGS-84 Z coordinate

The satellite clock offset can be estimated with the following equation:

$$\Delta t = a_0 + a_1(t - t_{oe}) + a_2(t - t_{oe})^2$$

3.1.2 Observation Equations of the Pseudoranges

Having calculated the satellite positions and clock offsets using the parameters transmitted from satellite navigation messages, the observation equation of a pseudorange can then be written as a function of the user coordinates and the receiver clock offset as follows:

$$\mathbf{d} = \mathbf{f}(\mathbf{x}) + \mathbf{v}$$

or in more detail:

$$
\begin{bmatrix} d_1 \\ d_2 \\ \vdots \\ d_n \end{bmatrix}
=
\begin{bmatrix}
\sqrt{(x_1 - x)^2 + (y_1 - y)^2 + (z_1 - z)^2} + \tau \\
\sqrt{(x_2 - x)^2 + (y_2 - y)^2 + (z_2 - z)^2} + \tau \\
\vdots \\
\sqrt{(x_n - x)^2 + (y_n - y)^2 + (z_n - z)^2} + \tau
\end{bmatrix}
+
\begin{bmatrix} v_1 \\ v_2 \\ \vdots \\ v_n \end{bmatrix}
\tag{3.1}
$$

where $[x_i, y_i, z_i]$ is a coordinate vector of the position of the i-th satellite, $\mathbf{d} = [d_1, d_2, \ldots d_n]$ is a vector of the pseudorange measurements to n satellites ($n \geq 4$), while $\mathbf{x} = [x, y, z, \tau]^\mathrm{T}$ is the a parameter vector containing the user coordinates x, y, z and the receiver clock offset τ in meters. $\mathbf{v} = [v_1, v_2, \ldots v_n]^\mathrm{T}$ is a vector of measurement noise in meters.

Because there are typically more than four GNSS satellites in view, (3.1) typically represents an overdetermined system of equations. A common approach to use in such cases is the least squares estimate. The least squares estimate of (3.1) can be written as:

$$\hat{\mathbf{x}} = \arg \min_{\mathbf{x}} \left([\mathbf{d} - \mathbf{f}(\mathbf{x})]^T [\mathbf{d} - \mathbf{f}(\mathbf{x})] \right) \tag{3.2}$$

This is a nonlinear system of equations that can be solved either using a Taylor series expansion to linearize the equations, or using a closed-form approach. The linearization approach will be described in Section 3.1.3, while the closed-form approach will be discussed in Section 3.1.4. Last, the parameter vector $\mathbf{x}$ is often estimated with a Kalman filter, which will be covered in Section 3.1.5.

3.1.3 Linear Least Squares Estimate Based on the Taylor Series Expansion

The Taylor series expansion is a common approach for approximating a nonlinear function, as follows:

$$f(\mathbf{x}) = f(\mathbf{x}_0) + \left.\frac{\partial f}{\partial \mathbf{x}}\right|_{\mathbf{x}=\mathbf{x}_0} d\mathbf{x} + \left.\frac{\partial^2 f}{\partial \mathbf{x}^2}\right|_{\mathbf{x}=\mathbf{x}_0} d\mathbf{x}^2 + \ldots \tag{3.3}$$

where $\mathbf{x}_0$ is the initial user position from which the Taylor series is expanded. When the terms of the second order and higher are ignored, the nonlinear function is approximated by the following linear function:

$$f(\mathbf{x}) = f(\mathbf{x}_0) + \left.\frac{\partial f}{\partial \mathbf{x}}\right|_{\mathbf{x}=\mathbf{x}_0} d\mathbf{x}, \tag{3.4}$$

where

$$\mathbf{x}_0 = \begin{bmatrix} x_0 \\ y_0 \\ z_0 \\ \Delta t_0 \end{bmatrix}, \quad \left.\frac{\partial f}{\partial \mathbf{x}}\right|_{\mathbf{x}=\mathbf{x}_0} = \begin{bmatrix} \dfrac{\partial f}{\partial x} & \dfrac{\partial f}{\partial y} & \dfrac{\partial f}{\partial z} & \dfrac{\partial f}{\partial \tau} \end{bmatrix}_{\mathbf{x}=(x_0 \;\; y_0 \;\; z_0 \;\; \tau_0)}$$

$$d\mathbf{x} = \begin{bmatrix} x - x_0 \\ y - y_0 \\ z - z_0 \\ \tau - \tau_0 \end{bmatrix}.$$

Using (3.4) to replace the nonlinear functions in (3.1), we can express the pseudoranges as follows:

$$\begin{aligned} \mathbf{d} &= \mathbf{f}(\mathbf{x}) + \mathbf{v} \\ &= \mathbf{f}(\mathbf{x}_0) + \mathbf{H}d\mathbf{x} + \mathbf{v} \end{aligned} \tag{3.5}$$

where

$$\mathbf{f}(\mathbf{x}_0) = \begin{bmatrix} \sqrt{(x_1 - x_0)^2 + (y_1 - y_0)^2 + (z_1 - z_0)^2} + \tau_0 \\ \sqrt{(x_2 - x_0)^2 + (y_2 - y_0)^2 + (z_2 - z_0)^2} + \tau_0 \\ \vdots \\ \sqrt{(x_n - x_0)^2 + (y_n - y_0)^2 + (z_n - z_0)^2} + \tau_0 \end{bmatrix}$$

and

$$\mathbf{H} = \begin{bmatrix} \left.\dfrac{\partial f_1}{\partial x}\right|_{\mathbf{x}=\mathbf{x}_0} & \left.\dfrac{\partial f_1}{\partial y}\right|_{\mathbf{x}=\mathbf{x}_0} & \left.\dfrac{\partial f_1}{\partial z}\right|_{\mathbf{x}=\mathbf{x}_0} & \left.\dfrac{\partial f_1}{\partial \tau}\right|_{\mathbf{x}=\mathbf{x}_0} \\ \vdots & \vdots & \vdots & \vdots \\ \left.\dfrac{\partial f_n}{\partial x}\right|_{\mathbf{x}=\mathbf{x}_0} & \left.\dfrac{\partial f_n}{\partial y}\right|_{\mathbf{x}=\mathbf{x}_0} & \left.\dfrac{\partial f_n}{\partial z}\right|_{\mathbf{x}=\mathbf{x}_0} & \left.\dfrac{\partial f_n}{\partial \tau}\right|_{\mathbf{x}=\mathbf{x}_0} \end{bmatrix}_{n \times 4}.$$

The partial derivatives in matrix $\mathbf{H}$ and values of the functions at the initial point $\mathbf{f}(\mathbf{X}_0)$ can be expressed as:

$$\frac{\partial f_i}{\partial x} = \frac{x_0 - x_i}{d_0^i}$$

$$\frac{\partial f_i}{\partial y} = \frac{y_0 - y_i}{d_0^i}$$

$$\frac{\partial f_i}{\partial z} = \frac{z_0 - z_i}{d_0^i}$$

$$\frac{\partial f_i}{\partial \tau} = 1$$

$$f^i(\mathbf{x}_0) = \sqrt{(x_0 - x_i)^2 + (y_0 - y_i)^2 + (z_0 - z_i)^2} + \tau_0$$

Equation (3.5) represents a linear system of equations. The least squares solution of (3.5) can be written as:

$$\mathbf{dx} = (\mathbf{H}^T \mathbf{H})^{-1} \mathbf{H}^T (\mathbf{d} - \mathbf{f}(\mathbf{x}_0)) \qquad (3.6)$$

Finally, the estimate of the parameter vector, $\mathbf{x}$, is simply the sum of:

$$\mathbf{x} = \mathbf{x}_0 + \mathbf{dx} \qquad (3.7)$$

The covariance matrix of the estimated parameters (coordinates and clock offset) can be written as:

$$\mathbf{D}_{\mathbf{xx}} = \sigma_0^2 (\mathbf{H}^T \mathbf{H})^{-1} \qquad (3.8)$$

where σ_0^2 is the variance of the pseudorange observation.

A few iterations of (3.4) to (3.8) are typically needed in order to converge to a solution.

3.1.4 Closed-Form Least Squares Solution

The most significant advantage of the Taylor series expansion is that it can be applied to any nonlinear function representing a set of measurements. However, it has its disadvantages as well. For example, it requires an (approximate) initial position, and it needs multiple iterations (three to five times) to converge to a final solution. As such, it is not the most computationally efficient solution. For the above reasons, various closed-form solutions have been investigated and developed (e.g., by Bancroft [1], Krause [2], Abel and Chafee [3, 4], Hoshen [5], Nardi and Pachter [6], and Nguyen [7]).

One approach removes the nonlinearity of the pseudorange observations by taking the difference of the square of two pseudoranges. First, we rearrange the function for the pseudorange as follows:

$$
\begin{bmatrix} d_1 - \tau \\ d_2 - \tau \\ \vdots \\ d_n - \tau \end{bmatrix} = \begin{bmatrix} \sqrt{(x_1 - x)^2 + (y_1 - y)^2 + (z_1 - z)^2} \\ \sqrt{(x_2 - x)^2 + (y_2 - y)^2 + (z_2 - z)^2} \\ \vdots \\ \sqrt{(x_n - x)^2 + (y_n - y)^2 + (z_n - z)^2} \end{bmatrix} \tag{3.9}
$$

Taking the squares of both sides of (3.9), we have:

$$
\begin{bmatrix} d_1^2 - 2d_1\tau + \tau^2 \\ d_2^2 - 2d_2\tau + \tau^2 \\ \vdots \\ d_n^2 - 2d_n\tau + \tau^2 \end{bmatrix} = \begin{bmatrix} x^2 + y^2 + z^2 + x_1^2 + y_1^2 + z_1^2 - 2(x_1 x + y_1 y + z_1 z) \\ x^2 + y^2 + z^2 + x_2^2 + y_2^2 + z_2^2 - 2(x_2 x + y_2 y + z_2 z) \\ \vdots \\ x^2 + y^2 + z^2 + x_n^2 + y_n^2 + z_n^2 - 2(x_n x + y_n y + z_n z) \end{bmatrix} \tag{3.10}
$$

Taking the difference of the ith equation and the first equation in (3.10), the terms x^2, y^2, z^2, and τ^2 are canceled, and the system of equations becomes a system of linear equations as follows:

$$
\begin{bmatrix} d_2^2 - d_1^2 - 2d_{21}\tau \\ d_3^2 - d_1^2 - 2d_{31}\tau \\ \vdots \\ d_n^2 - d_1^2 - 2d_{n1}\tau \end{bmatrix} = \begin{bmatrix} x_2^2 + y_2^2 + z_2^2 - x_1^2 - y_1^2 - z_1^2 + 2x_{12}x + 2y_{12}y + 2z_{12}z \\ x_3^2 + y_3^2 + z_3^2 - x_1^2 - y_1^2 - z_1^2 + 2x_{13}x + 2y_{13}y + 2z_{13}z \\ \vdots \\ x_n^2 + y_n^2 + z_n^2 - x_1^2 - y_1^2 - z_1^2 + 2x_{1n}x + 2y_{1n}y + 2z_{1n}z) \end{bmatrix} \tag{3.11}
$$

where the notation $a_{ij} = a_i - a_j$ is used to conserve space (i.e., $d_{ij} = d_i - d_j$ and $x_{ij} = x_i - x_j$). By rearranging (3.11), the following system of linear equations can be obtained:

$$
\begin{bmatrix}
x_{12} & y_{12} & z_{12} & d_{21} \\
x_{13} & y_{13} & z_{13} & d_{31} \\
\vdots & \vdots & \vdots & \vdots \\
x_{1n} & y_{1n} & z_{1n} & d_{n1}
\end{bmatrix}
\begin{bmatrix}
x \\ y \\ z \\ \tau
\end{bmatrix}
= \frac{1}{2}
\begin{bmatrix}
x_1^2 + y_1^2 + z_1^2 - d_1^2 - x_2^2 - y_2^2 - z_2^2 + d_2^2 \\
x_1^2 + y_1^2 + z_1^2 - d_1^2 - x_3^2 - y_3^2 - z_3^2 + d_3^2 \\
\vdots \\
x_1^2 + y_1^2 + z_1^2 - d_1^2 - x_n^2 - y_n^2 - z_n^2 + d_n^2
\end{bmatrix}
\tag{3.12}
$$

This can be written more compactly in matrix and vector form as:

$$
\mathbf{H}\mathbf{x} = \mathbf{b}, \tag{3.13}
$$

where

$$
\mathbf{H} =
\begin{bmatrix}
x_{12} & y_{12} & z_{12} & d_{21} \\
x_{13} & y_{13} & z_{13} & d_{31} \\
\vdots & \vdots & \vdots & \vdots \\
x_{1n} & y_{1n} & z_{1n} & d_{n1}
\end{bmatrix}
$$

$$
\mathbf{x} =
\begin{bmatrix}
x \\ y \\ z \\ \tau
\end{bmatrix}
\quad \text{and}
$$

$$
\mathbf{b} = \frac{1}{2}
\begin{bmatrix}
x_1^2 + y_1^2 + z_1^2 - d_1^2 - x_2^2 - y_2^2 - z_2^2 + d_2^2 \\
x_1^2 + y_1^2 + z_1^2 - d_1^2 - x_3^2 - y_3^2 - z_3^2 + d_3^2 \\
\vdots \\
x_1^2 + y_1^2 + z_1^2 - d_1^2 - x_n^2 - y_n^2 - z_n^2 + d_n^2
\end{bmatrix}
$$

The positions of the n satellites are known from the broadcast navigation message, and the pseudorange measurements d_i are obtained by the receiver. In an overdetermined system (i.e., more than four visible satellites), the least squares solution of (3.13) can be obtained directly without any iteration with:

$$\mathbf{x} = (\mathbf{H}^T\mathbf{H})^{-1}\mathbf{H}^T\mathbf{b} \tag{3.14}$$

Similar to the Taylor's series expansion solution explained in Section 3.1.3, the covariance matrix of the estimated parameters can be estimated with:

$$\mathbf{D_{xx}} = \sigma_0^2(\mathbf{H^TH})^{-1} \tag{3.15}$$

The closed-form solution given in (3.14) and (3.15) is very computationally efficient because no iteration is needed to find the least squares solution. However, there is a significant disadvantage for this solution: It requires five pseudoranges instead of four to find a solution for four parameters. This is not a problem for most cases when we have an open view to sky, especially when there will be more GNSSs operational in the near future.

3.1.5 The Kalman Filter Solution

The Kalman filter is the common approach for parameter estimation that has been adopted by many fields of engineering. This is also the case for GNSS positioning in mobile devices. The Kalman filter applies a model of user dynamics into the location estimation and, as a result, gives a smoother user trajectory estimate. Least squares solutions are often used only to find the initial state of the Kalman filter.

For a Kalman filter solution, we need to define 1) a state vector, 2) a system process model, and 3) an observation model. For mobile positioning, location estimation is typically carried out in the local geodetic coordinate system as defined in Chapter 2.1.3. The following simple example illustrates a simple Kalman filter solution for GNSS positioning in a mobile device.

Start with a state vector, which describes the state of the mobile device:

$$\mathbf{x} = \left[n, v_n, e, v_e, h, v_h, \tau, \dot{\tau}\right]^T \tag{3.16}$$

where

$\quad n,e,h$ are local geodetic coordinates of the north, east, and up components;

$\quad v_n, v_e, v_h$ are the corresponding velocity components;

$\quad \tau$ is the receiver clock offset in terms of meters;

$\quad \dot{\tau}$ is velocity component of the receiver clock offset.

We adopt a system process model, which describes how the state of the mobile device changes from one time epoch to the next:

$$\mathbf{x}_{k+1} = \mathbf{\Phi}_k \mathbf{x}_k + \mathbf{w}_k, \quad \mathbf{w}_k \sim N(0, \mathbf{Q})$$

where

$\mathbf{\Phi}$	is the state transition matrix;
$\mathbf{w}$	is the vector of white noise;
$\mathbf{Q}$	is the process noise matrix;
k	is the index of the time epoch.

The observation model, which describes how the measurements (e.g., pseudoranges) relate to the parameters of the state vector, is assumed to be linear as follows:

$$\mathbf{d}_k = \mathbf{H}\mathbf{x}_k + \mathbf{v}_k \quad \text{with } \mathbf{H} = \frac{\partial f(\mathbf{x}_k)}{\partial \mathbf{x}_k}, \quad \mathbf{v}_k \sim N(0, \mathbf{R}) \tag{3.17}$$

where

$\mathbf{d}$	is a vector of observations;
$\mathbf{v}$	is a vector of measurement errors;
$\mathbf{H}$	is the design matrix of the observation equations;
$\mathbf{R}$	is the covariance matrix of the pseudorange measurements, which can be a diagonal matrix with equal variance for all measurements.

By initiating the *Kalman filter* using a least squares solution for the initial epoch with:

$$\mathbf{x}_0^- = (\mathbf{H}_0^T \mathbf{H}_0)^{-1} \mathbf{H}_0^T \mathbf{d}_0$$
$$\mathbf{P}_0^- = \sigma_0^2 (\mathbf{H}_0^T \mathbf{H}_0)^{-1}$$

where σ_0^2 is the variance of the pseudorange observations and $\mathbf{P}_0^-$ is the covariance matrix of the estimated parameters. The Kalman filter update steps can be calculated as:

$$\mathbf{K}_k = \mathbf{P}_k^- \mathbf{H}_k^T \left[\mathbf{H}_k \mathbf{P}_k^- \mathbf{H}_k^T + \mathbf{R}_k \right]^{-1}$$
$$\mathbf{x}_k^+ = \mathbf{x}_k^- + \mathbf{K}_k \left[\mathbf{d}_k - \mathbf{H}_k \mathbf{x}_k^- \right] \tag{3.18}$$
$$\mathbf{P}_k^+ = \mathbf{P}_k^- - \mathbf{K}_k \mathbf{H}_k \mathbf{P}_k^-$$

where $\mathbf{K}$ is the gain matrix. The projection operation can then be calculated as:

$$x_{k+1}^- = \Phi_k x_k^+$$

$$P_{k+1}^- = \Phi_k P_k^+ \Phi_k^T + Q_k \tag{3.19}$$

In order to obtain the state transition matrix Φ_k, the dynamic process needs to be defined [8] as:

$$\dot{x} = Fx + u$$

where

$$F = \begin{bmatrix} 0 & 1 & 0 & 0 & 0 & 0 & 0 & 0 \\ 0 & 0 & 0 & 0 & 0 & 0 & 0 & 0 \\ 0 & 0 & 0 & 1 & 0 & 0 & 0 & 0 \\ 0 & 0 & 0 & 0 & 0 & 0 & 0 & 0 \\ 0 & 0 & 0 & 0 & 0 & 1 & 0 & 0 \\ 0 & 0 & 0 & 0 & 0 & 0 & 0 & 0 \\ 0 & 0 & 0 & 0 & 0 & 0 & 0 & 1 \\ 0 & 0 & 0 & 0 & 0 & 0 & 0 & 0 \end{bmatrix}, \text{ and } u = \begin{bmatrix} 0 \\ u_n \\ 0 \\ u_e \\ 0 \\ u_h \\ u_\tau \\ u_{\dot{\tau}} \end{bmatrix} \tag{3.20}$$

The state transition matrix Φ_k can then be derived as

$$\Phi_k = e^{F\Delta t}$$

$$= I + F\Delta t \tag{3.21}$$

Applying (3.20) to (3.21), we have

$$\Phi_k = \begin{bmatrix} 1 & \Delta t & 0 & 0 & 0 & 0 & 0 & 0 \\ 0 & 1 & 0 & 0 & 0 & 0 & 0 & 0 \\ 0 & 0 & 1 & \Delta t & 0 & 0 & 0 & 0 \\ 0 & 0 & 0 & 1 & 0 & 0 & 0 & 0 \\ 0 & 0 & 0 & 0 & 1 & \Delta t & 0 & 0 \\ 0 & 0 & 0 & 0 & 0 & 1 & 0 & 0 \\ 0 & 0 & 0 & 0 & 0 & 0 & 1 & \Delta t \\ 0 & 0 & 0 & 0 & 0 & 0 & 0 & 1 \end{bmatrix}$$

where Δt is the time span between the epochs $k - 1$ and k. The processing noise matrix Q_k can be derived as [8]:

$$\mathbf{Q}_k = E\left[\mathbf{w}_k \mathbf{w}_k^T\right] = \begin{vmatrix} \mathbf{Q}_n & 0 & 0 & 0 \\ 0 & \mathbf{Q}_e & 0 & 0 \\ 0 & 0 & \mathbf{Q}_h & 0 \\ 0 & 0 & 0 & \mathbf{Q}_\tau \end{vmatrix} \tag{3.22}$$

where

$$\mathbf{Q}_n = \begin{bmatrix} \dfrac{s_n \Delta t^3}{3} & \dfrac{s_n \Delta t^2}{2} \\ \dfrac{s_n \Delta t^2}{2} & s_n \Delta t \end{bmatrix} \tag{3.23}$$

$$\mathbf{Q}_e = \begin{bmatrix} \dfrac{s_e \Delta t^3}{3} & \dfrac{s_e \Delta t^2}{2} \\ \dfrac{s_e \Delta t^2}{2} & s_e \Delta t \end{bmatrix} \tag{3.24}$$

$$\mathbf{Q}_h = \begin{bmatrix} \dfrac{s_h \Delta t^3}{3} & \dfrac{s_h \Delta t^2}{2} \\ \dfrac{s_h \Delta t^2}{2} & s_h \Delta t \end{bmatrix} \tag{3.25}$$

$$\mathbf{Q}_\tau = \begin{bmatrix} s_b \Delta t + \dfrac{1}{3} s_f \Delta t^3 & \dfrac{1}{2} s_f \Delta t^2 \\ \dfrac{1}{2} s_f \Delta t^2 & s_f \Delta t \end{bmatrix} \tag{3.26}$$

Here s_n, s_e, s_h are the power spectral densities associated with the velocity components of the north, east, and height coordinate components, respectively, while s_b and s_f are the power spectral densities associated with the clock offset and clock drift components, respectively.

3.2 DIFFERENTIAL GNSS

As shown in (3.8) and (3.15), the covariance matrix of the estimated parameters is related to two issues: the variance of the pseudorange observation σ_0^2 and the satellite geometry that define the design matrix **H**. For the code-based positioning solution using pseudoranges, the error sources include: satellite clock error, satellite orbit error, ionospheric delay, tropospheric delay, receiver noise, and multipath. These error sources are summarized in Table 3.3 [9].

Table 3.3
Error Sources for Pseudorange Based GNSS Positioning Solution

Error Sources	*Error (m)*
Satellite clock	1.1
Satellite orbit	0.8
Ionospheric delay	7.0
Troposheric delay	0.2
Receiver noise	0.1
Multipath	0.2

Many of these errors are highly correlated over space and time. For example, the difference of the vertical ionospheric delays within a small area (e.g., less than 100 km) is typically insignificant (< 0.5m) under an undisturbed ionospheric condition. The spatial and time correlation gives us an opportunity to eliminate or at least to mitigate the impact of these errors by determining error corrections at one or more known locations and delivering these error corrections to the users in the surroundings. This approach is called DGNSS. Figure 3.1 shows the basic concept of a DGNSS. The operation procedure of a DGNSS includes the following steps:

- Estimating the errors at reference stations;
- Broadcasting the errors, in the form of correction messages, to the users in the surroundings of the reference stations;
- Applying the corrections to the pseudorange measurements before using them for positioning.

The error estimation is carried out at the reference stations. It is achieved by simply subtracting the calculated ranges d_c from the observed pseudoranges d_o as follows:

$$e_i = d_o - d_c$$

where

$$d_c = \sqrt{(x_i - x)^2 + (y_i - y)^2 + (z_i - z)^2}$$

(3.27)

In (3.27), (x_i, y_i, z_i) are the coordinates of the corresponding satellite, which are calculated from the ephemeris as explained in Section 3.1.1, while (x, y, z) are the coordinates of the reference station, which are known *a priori* to a high level of accuracy (e.g., a few centimeters).

Having estimated the errors for all visible satellites in view, these errors are then coded in the form of correction messages before being broadcasted to the mobile users in the surroundings. Obviously a data communication link is needed for transmitting the error corrections to the mobile users. It can be a dedicated radio communication link or a wireless Internet connection.

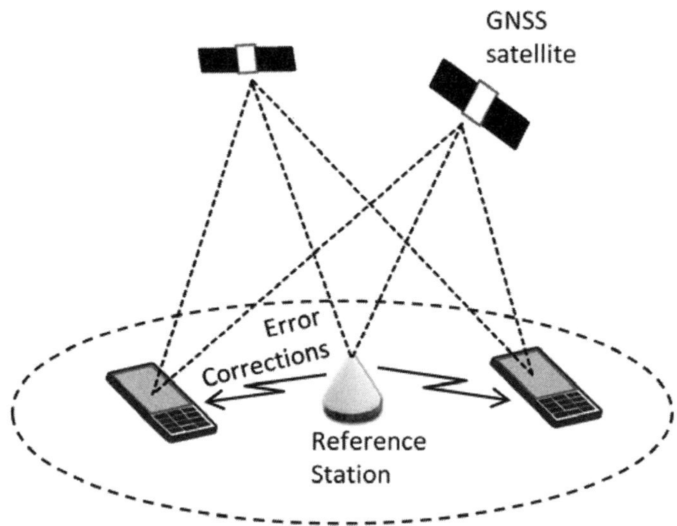

Figure 3.1 The concept of a differential GNSS solution.

After receiving the error corrections over a data communication link, the mobile receiver decodes the correction messages and applies these error corrections to pseudorange measurements before applying them for position estimation using one of the positioning algorithms explained in Sections 3.1.3–3.1.5.

By taking advantage of the high spatial and time correlation of the error sources, DGNSS mitigates significantly the impacts of various errors to the final position estimate.

The minimum configuration of the DGNSS system can theoretically consist of only one reference station. The effective distance of the error corrections is about 150 km from the reference station. To enlarge the service area, a network of reference stations (e.g., a nationwide network), is typically deployed to offer a nationwide DGNSS service.

3.3 SBAS

SBAS is another type of service that can improve the positioning accuracy by applying error corrections to the pseudoranges. It was originally designed for aviation applications to support precision approach of aircraft. Differing from the DGNSS, SBAS delivers the postcorrected error bounds of the pseudorange measurements in addition to the error corrections; therefore, the end users can estimate the error bounds of the position estimates with a certain confidence level. This is important for SoL applications, such as precision approach of aircrafts. The error bounds (or protection levels) of the position estimate are used for integrity assessment, which will alert the end users whenever the error bound of a position estimate is larger than the threshold for alarm. The SBAS service is also attractive to non-SoL applications, such as toll road pricing, legal applications, car insurance pricing, and precise agriculture. Because of the large application scope and the fact that it is easy to implement, most GNSS chips, including those used in smartphones, have implemented the SBAS functions. No additional hardware is needed to implement SBAS support; only a software update is needed. Of course, sufficient computational resources are needed to decode the 1-Hz SBAS messages and to apply 1) the error corrections for positioning, and 2) error bound information for integrity assessment. It is relevant to address this topic briefly in this chapter.

The architecture of a SBAS consists of three segments: the space segment, the ground segment, and the user segment (as in many GNSSs), as shown in Figure 3.2.

The space segment typically includes a few GEO satellites that broadcast the SBAS signal in space (SBAS messages) in the GPS L1 frequency. Furthermore, the GEO satellites offer additional ranging measurements that can be used for positioning as well.

The ground segment consists of a network of reference stations installed at presurveyed locations from where measurements are collected in real time and used for generating the SBAS signal in space, which includes the error corrections and the integrity information needed for calculating the post corrected error bounds of the pseudoranges. In addition to a network of reference stations, data processing facilities are needed in order to estimate the error corrections and integrity information. Furthermore, a few uplink stations are required in the ground segment in order to upload the SBAS messages to GEO satellites.

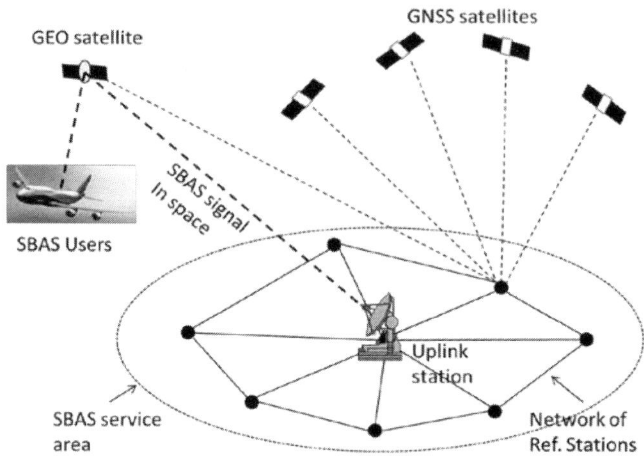

Figure 3.2 The segments of the SBAS.

The user segment is a GNSS receiver that is capable of receiving the SBAS signal in space, decoding the SBAS messages, and utilizing the information for positioning and integrity assessment.

The GEO satellite transmits one SBAS message per second. Each SBAS message contains 212 bits of data field [10] that contain the error corrections and integrity information as shown in Figure 3.3.

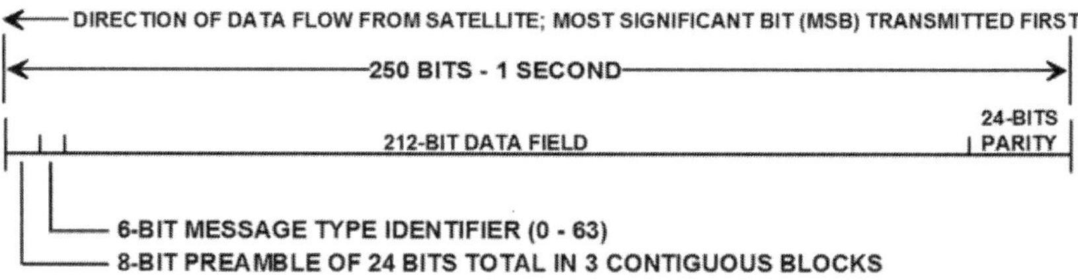

Figure 3.3 The format of a SBAS message.

The error corrections broadcasted in the SBAS signal in space include the satellite clock corrections, satellite orbit corrections, and the ionospheric corrections. The satellite clock corrections are further divided into the fast corrections and long-term corrections. The fast corrections were designed to remove the rapid change errors introduced by the selective availability (SA) function of GPS, which was turned off on May 1, 2000. Therefore, this part is primarily an artifact of legacy. The long-term corrections to the satellite clocks and orbit corrections, however, remain important, even though they do not change very rapidly. The vertical ionospheric delays are transmitted for the five-by-five degree grid points within the service coverage areas [10]. Corrections to the tropospheric delays are not included in the SBAS signal in space, but rather they are calculated based on an empirical model using the user's approximate position and day of the year as input parameters.

The least squares solution is recommended for SBAS positioning, in order to assure fast system reaction to the case of malfunction of the GNSS system [10]. Complex positioning solutions, such as a Kalman filter, might delay the system response for such situations.

The following weighted least squares solution is adopted for SBAS positioning and integrity assessment. It starts with a linear observation equation:

$$\begin{aligned} \mathbf{d} &= \mathbf{f}(\mathbf{x}) + \mathbf{v} \\ &= \mathbf{f}(\mathbf{x}_0) + \mathbf{H}\,\mathbf{dx} + \mathbf{v}, \qquad \mathbf{v} \sim N(0, \mathbf{P}^{-1}) \end{aligned} \tag{3.28}$$

or

$$\mathbf{l} = \mathbf{H}\,\mathbf{dx} + \mathbf{v} \quad \text{with } \mathbf{l} = \mathbf{d} - \mathbf{f}(\mathbf{x}_0),$$

where

$$\mathbf{dx} = \begin{bmatrix} de \\ dn \\ dh \\ d\tau \end{bmatrix} = \begin{bmatrix} e - e_0 \\ n - n_0 \\ h - h_0 \\ \tau - \tau_0 \end{bmatrix}.$$

Here (e, n, h) are the east, north, and height coordinates, respectively, τ is the clock offset, and $\mathbf{v}$ is a vector of measurement errors. The design matrix $\mathbf{H}$ can be written as:

$$\mathbf{H} = \begin{bmatrix} \cos E_1 \sin A_1 & \cos E_1 \cos A_1 & \sin E_1 & 1 \\ \cos E_2 \sin A_2 & \cos E_2 \cos A_2 & \sin E_2 & 1 \\ \vdots & \vdots & \vdots & \vdots \\ \cos E_m \sin A_m & \cos E_m \cos A_m & \sin E_m & 1 \end{bmatrix}$$

where E and A are the elevation angle and azimuth of the corresponding satellite, respectively, while m is the number of satellites used in the positioning solution. The weight matrix $\mathbf{W}$ is defined as:

$$W = \begin{bmatrix} \sigma_1^2 & 0 & \cdots & 0 \\ 0 & \sigma_2^2 & & \vdots \\ \vdots & & \ddots & 0 \\ 0 & \cdots & 0 & \sigma_m^2 \end{bmatrix}^{-1}$$

where

$$\sigma_i^2 = \sigma_{i,flt}^2 + \sigma_{i,UIRE}^2 + \sigma_{i,air}^2 + \sigma_{i,tropo}^2 \tag{3.29}$$

The terms in (3.29) are defined as follows:

- $\sigma_{i,flt}^2$ is the variance of the fast and long-term corrections obtained from the SBAS signal in space;

- $\sigma_{i,UIRE}^2$ is the variance of the ionospheric delay corrections obtained from the SBAS signal in space;

- $\sigma_{i,air}^2$ is the variance of the airborne receiver errors obtained from a model;

- $\sigma_{i,tropo}^2$ is the variance of the tropospheric delays corrections obtained from a model [10].

Details on how to estimate the variance of the postcorrected pseudorange measurements can be found from [10]. The least squares solution of (3.28) can be written as:

$$dx = (H^T W H)^{-1} H^T W l$$
$$x = x_0 + dx$$

The covariance of the estimated parameters is:

$$D_{enh} = (H^T W H)^{-1} = \begin{bmatrix} \sigma_e^2 & \sigma_{en} & \sigma_{eh} & \sigma_{e\tau} \\ \sigma_{ne} & \sigma_n^2 & \sigma_{nh} & \sigma_{n\tau} \\ \sigma_{he} & \sigma_{hn} & \sigma_h^2 & \sigma_{h\tau} \\ \sigma_{\tau e} & \sigma_{\tau n} & \sigma_{\tau h} & \sigma_\tau^2 \end{bmatrix}.$$

Having estimated the covariance matrices of the estimated parameters, the integrity assessment can then be performed by estimating the vertical protection level

$$VPL = k_v \cdot \sigma_h$$

and the horizontal protection level

$$HPL = k_h \cdot \sigma_{major}$$
$$= k_h \cdot \sqrt{\frac{\sigma_e^2 + \sigma_n^2}{2} + \sqrt{\left(\frac{\sigma_e^2 - \sigma_n^2}{2}\right) + \sigma_{en}^2}}$$

where k_v and k_h are constants derived from the integrity requirements of the system. For example, for precision approach $k_v = 5.33$ and $k_h = 6.0$ were selected to bound the vertical and horizontal positions at significance levels of 10^{-7} and 2×10^{-9}, respectively, assuming that the error is characterized by a normal distribution [10].

Both horizontal and vertical protection levels are evaluated in real time in the user receiver for each second. Whenever the protection levels are larger than the alarm limits, the user will be alerted and the estimated coordinates should not be used for the corresponding operations.

3.4 A-GNSS

Both differential GNSS service and SBAS service offer error corrections to the end users to improve the positioning accuracy. However, these systems work only under the condition that the receiver is capable of tracking the satellites in view and that the ephemerides of the corresponding satellites are available in the user receivers. In order to obtain a complete ephemeris from a cold start, a receiver must continuously track a satellite for a minimum of 18 seconds without any interruption during signal reception. This is not a problem when an open view to the sky is available. However, it is very difficult to obtain an ephemeris if a user turns on the receiver in an urban canyon, from where it is hard to maintain uninterrupted tracking for more than 18 seconds because of the blockages of the high-rise buildings. For this reason, it sometimes takes tens of minutes to get the first position for mobile users who are moving around canyon cities. This is the major reason why assisted GNSS is needed. It can assist mobile devices in speeding up the TTFF process. A-GNSS is accomplished by delivering the satellite ephemerides and other assistance data over a wireless data connection. Therefore, there is no need for the receiver to acquire the ephemerides from the satellites directly and wait for at least 18 seconds to obtain the first position. Chen

[11] gives a very good introduction to this topic. More details can also be found from Diggelen [12].

Figure 3.4 shows the architecture of an A-GNSS service. It consists of a network of GNSS reference stations from which open views to sky are available, an A-GNSS server, a data communication link, and the A-GNSS–enabled receiver.

The reference stations operate continuously, where ephemerides of all satellites in view are continuously received and forwarded to the A-GNSS server, from which the ephemerides and other assistance data are delivered to the mobile users via a wireless data communication link.

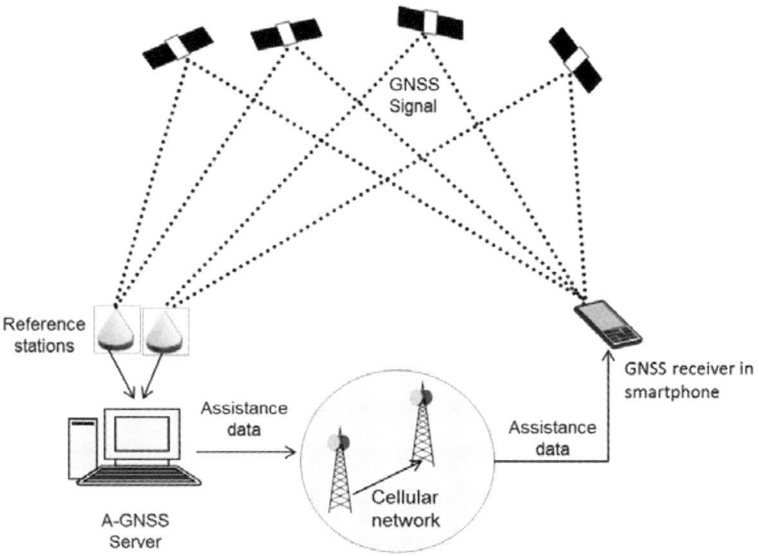

Figure 3.4 Architecture of an A-GNSS service.

A-GNSS also improves the TTFF in other ways. In order to estimate the user's position, pseudoranges to at least four satellites are needed in addition to satellite ephemerides, which are used for calculating the satellite positions. Therefore, fast signal acquisition is also needed for achieving a short TTFF (e.g., a few seconds). Signal acquisition is a process of searching a two-dimensional space with Doppler shift in one dimension, and code-delay in another dimension [11]. The search ranges are ±6 kHz for Doppler shift and 1,024 chips for code-delay. By knowing the approximate position of the mobile phone, the ephemerides of the satellites (for calculating the satellite positions and velocities) and a reference frequency obtained from the cellular network, it is possible to narrow down the searching space to ±0.6 kHz in the frequency dimension and 44 chips in the code-delay dimension [11]. The shrinkage of the searching space makes acquisition of the GNSS signals faster.

Both the satellite ephemerides and approximate user positions, which are determined by the cellular network based on the Cell-ID solution, are transmitted to the mobile users via the A-GNSS server as assistance data, while the reference frequency with an accuracy of ±0.1 ppm can be obtained directly from the signals of the cellular network [12].

In addition to speeding up the TTFF, the A-GNSS can also enhance the sensitivity of signal acquisition and tracking. This is achieved by providing the bit sequence of the navigation data in the assistance data, so that the receiver can remove the navigation data bits from the received signals. The removal of the navigation data bits from the received signal makes it possible to extend the length of coherent integration beyond the limitation of 20 ms. The signal tracking and acquisition sensitivity is then enhanced with the longer coherent integration length [11]. The A-GNSS server can obtain the bit sequences of the navigation data from the reference stations.

Unlike differential GNSS and SBAS services, A-GNSS services do not provide any error corrections to improve the positioning accuracy, but rather provide the satellite ephemerides and other assistance data to speed up the TTFF process and enhance the signal acquisition and tracking sensitivity. These enhancements are essential for positioning in urban canyons where the vast majority of smartphone users are located.

3.5 SUMMARY

This chapter starts with an introduction to the basic principles and algorithms for the standalone GNSS positioning solutions in mobile devices. It then introduces two solutions, the DGNSS solution and the solution based on SBAS, which provide error corrections to end users to enhance the positioning accuracy. The chapter concludes by presenting the A-GNSS service that delivers assistance data to the end users over a wireless network, for speeding up the TTFF process and enhancing the signal acquisition and tracking sensitivity. The A-GNSS service is an essential service for GNSS positioning, especially for urban environments where open views to the sky are typically not available.

References

[1]   Bancroft, S., "An algebraic solution of the GPS equations," *IEEE Transactions on Aerospace and Electronic Systems*, 21 (1), 1985, pp. 56-59.

[2]   Krause, L. O., "A direct solution to GPS type navigation equations," *IEEE Transactions on Aerospace and Electronic Systems*, 23(2), 1987, pp. 223-232.

[3]   Abel, J., and J. Chaffee, "Existence and uniqueness of GPS solutions," *IEEE Transactions on Aerospace and Electronic Systems*, 27(6), 1991, pp. 952-956.

[4]   Chaffee, J., and J. Abel, "On the exact solutions of pseudorange equations," *IEEE Transactions on Aerospace and Electronic Systems*, 30(4), 1994, pp. 1021-1030.

[5]   Hoshen, J., "The GPS equations and the problem of Apollonius," *IEEE Transactions on Aerospace and Electronic Systems*, 32(3), 1996, pp. 1116-1124.

[6]   Nardi, S., and M., Pachter, "GPS estimation algorithm using stochastic modeling," *Proc. the Conference on Decision and Control,* Tampa, FL, December, 1998.

[7]   Nguyen, T., "Efficient GPS Position Determination Algorithms," Ph.D Thesis, Airforce Institute of Technology. AFIT/DS/ENG/07-09.

[8]   Brown, R. G., and P. Hwang, *Introduction to Random Signals and Applied Kalman Filtering*, New York, John Wiley & Sons, 1997.

[9]   Kaplan, E. D., and C. J. Hegarty (eds). *Understanding GPS, Principles and Applications.* Second Edition, Norwood, MA: Artech House, 2006.

[10] RTCA, Minimum Operational Performance Standards for Global Positioning System/Wide Area Augmentation System Airborne Equipment. RTCA/DO-229C, 2001.

[11] Chen, R., "Assisted GNSS in smart phones," In *Ubiquitous Positioning and Mobile Location-Based Services in Smart Phones*, pp. 32–43, R. Chen (ed), Pennsylvania: IGI-Global, 2012.

[12] Diggelen, F. V., *A-GPS, Assisted GPS, GNSS and SBAS*, Norwood, MA: Artech House, 2009.

Chapter 4

Wireless Positioning in Mobile Devices

Although GNSS technologies, especially the GPS, have been widely adopted as the default positioning technology for most location-based services (LBS), it is still not possible to achieve a ubiquitous positioning solution with this single technology. Problems are especially apparent in urban areas or indoor environments, where the weak signals of GNSS do not penetrate building materials or suffer from severe multipath effects. Other sensors and radio signals are commonly adopted, in order to augment the positioning solution and enhance its availability in all environments. This chapter addresses the topic of using radio frequency (RF) measurements other than GNSS signals for mobile positioning.

Communication is the main function of most mobile devices, and there are various forms of wireless connectivity available in mobile devices for facilitating the communication function. These typically employ RF signals through wireless radio access networks (e.g., 3G, 4G, WLAN networks, and Bluetooth networks).[13] Although these RF signals were not originally designed for positioning purposes, they exhibit many properties that are spatially correlated, such as radio signal strength, thus providing an opportunity to adopt these RF signals for positioning, especially for indoor environments. This is why these RF signals are often called "signals of opportunity" when viewed from the perspective of mobile positioning.

There are various approaches available for mobile positioning using these RF signals. These approaches are based on different RF observables such as radio signal coverage; radio signal strength; direction of signal arrival; range (via ToA); and range difference (via TDoA). Different positioning algorithms have been developed for utilizing these RF observables for locating mobile devices. These algorithms are adopted for: 1) positioning solutions based on radio signal coverage area; 2) positioning solutions based on radio signal pattern; 3) positioning solutions based on range and range difference; and 4) positioning solutions based on direction of signal arrival. Positioning accuracy of these

---

[13] 3G and 4G are abbreviations for the third generation and fourth generation of mobile telecommunication technologies, standardized primarily by the 3rd Generation Partnership Project (3GPP) and regulated by the International Telecommunications Union (ITU).

algorithms varies from a few meters indoors to tens of kilometers outdoors, depending on the observables used for positioning, the density of the RF transmitters in the radio infrastructures, and the positioning environments.

These algorithms are applicable to different types of wireless radio infrastructures including 3G or 4G networks, WLAN access points, Bluetooth access points, and RFID tags.

4.1 POSITIONING BASED ON RADIO SIGNAL COVERAGE AREA

The signal coverage area approach is the simplest of the RF positioning methods that will be described below. It is based on the fact that when a mobile device makes a connection to a RF transmitter, such as a base station of a cellular network or an access point of a WLAN or a Bluetooth network, it is identified as within the radio coverage area of this RF transmitter or within the effective range of the RF transmitter. For positioning purposes, the locations of these RF transmitters are typically known; therefore, the mobile device can be easily located as "within the radio coverage area of the corresponding RF transmitter" as shown in Figure 4.1 [1, 2]. This is called the Cell-ID approach as explained in Section 2.1.1.1.

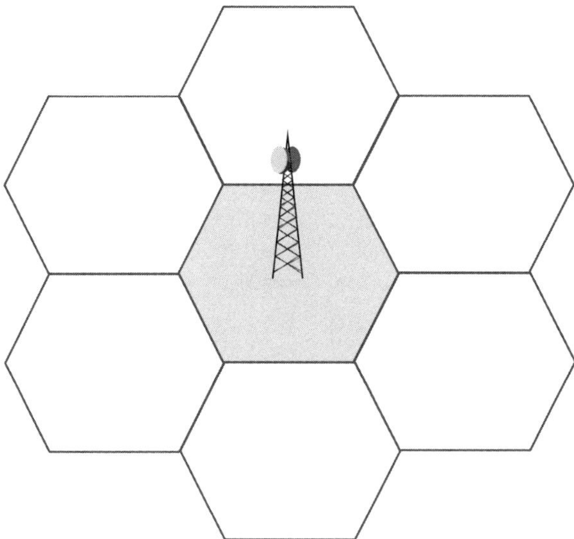

Figure 4.1 Mobile positioning based on the radio signal coverage. The location of the RF transmitter is the output of the positioning solution, while the maximum effective range of the RF signal is the accuracy of the positioning solution.

In terms of coordinates, the location of the transmitter will be taken as the position of the mobile device. The positioning accuracy will be equal to the maximum effective range of the transmitter. The positioning error ranges from a few tens of meters in a dense WLAN network indoors to tens of kilometers in rural areas, where cellular networks are sparse.

If additional information is known about the RF signal, additional constraints can be applied to the positioning solutions, in order to achieve a better positioning accuracy. Typical constraints can be a distance (i.e., range), a DoA, or a combination of both [2, 3]. In effect, these constraints further reduce the potential area in which the mobile device could be. This implies that the positioning accuracy is improved. When a range constraint is used, the signal coverage area method becomes similar to the range approach. Likewise when a DoA constraint is used, it becomes similar to the DoA approach. The main difference lies in the fact that the signal coverage area approach typically only uses information from a single RF transmitter at any given time, where the other approaches combine information simultaneously from multiple transmitters.

As discussed in Section 2.2.1, a distance can be estimated from a ToA measurement between the transmitter and the receiver [4] [see (2.11)], or estimated from an observed signal strength via a path loss model [e.g., the one given in (2.12)] [5, 6]. In addition to the distance d, the error Δd in the distance measurement should be estimated. Combining the distance d and the corresponding measurement error Δd, the potential positions of the mobile device are then reduced from the entire signal coverage area to a ring, whose center is at the position of the transmitter and whose radius is reduced to d. The width of the ring is defined by the error in the distance measurement Δd, as shown in Figure 4.2.

As mentioned in Section 2.2.4, the DoA of the RF signal from the transmitter can be estimated either by a directional antenna or an antenna array [7, 8]. This DoA measurement is used to define a line between the transmitter and the receiver. For a solution based on a direction constraint, the area of the potential positions of the mobile device is narrowed to an angular section extending from the transmitter to the receiver, whose central line follows the estimated DoA. The angular width of this section is defined by the uncertainty of the estimated DoA, as shown in Figure 4.3 [3].

In case both distance and DoA information of the received RF signal are available, the area of the potential positions of the mobile device can be further reduced to an area defined by the intersection of a ring (as in Figure 4.2) and angular section (as in Figure 4.3). This intersecting region, known as an annular sector, represents a significantly smaller area of potential locations of the mobile device compared to the entire signal coverage area—thus the position accuracy is higher using this more advanced approach.

The signal coverage area solutions with constraints are called enhanced-Cell-ID solutions as mentioned in Section 2.1.1.1.

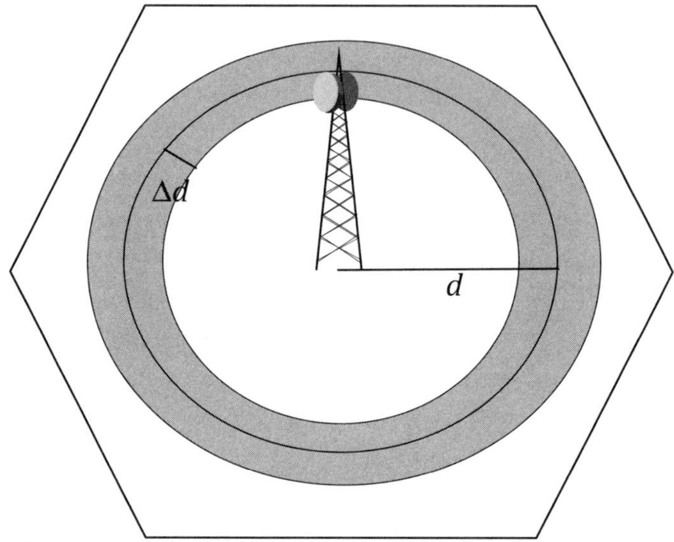

Figure 4.2 A distance-constrained signal coverage area solution. The gray area is area of potential position of the mobile device.

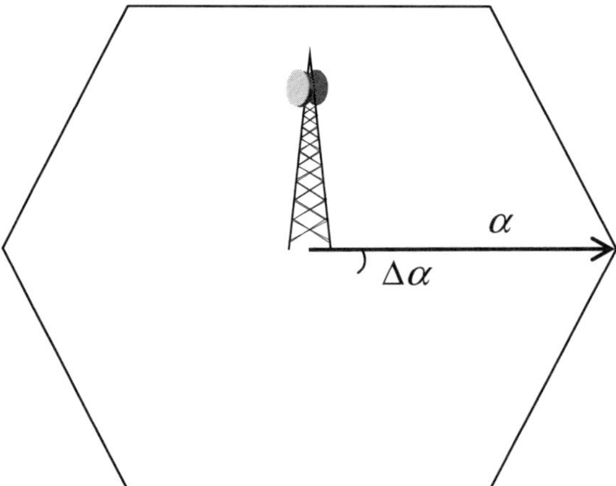

Figure 4.3 A direction-constrained signal coverage area solution. The gray area is the area of potential position of the mobile device.

4.2 POSITIONING BASED ON RADIO SIGNAL PATTERN

In many circumstances, a mobile device can measure RF signals not just from a single transmitter but from multiple ones. In these cases, the combination of measured signals forms a radio signal pattern, which is another important observable adopted for mobile device positioning. The signal pattern is typically comprised of a vector of signal identifiers (such as a Cell-ID or MAC address) and corresponding signal strength measurements.

Fingerprinting is the most common approach for mobile device positioning using RF signal patterns. This approach is known as fingerprinting because it is analogous to the way fingerprints are used to identify people. In this analogy, signal patterns serve as the "fingerprints," and the locations corresponding to the signal patterns are the identities to be recognized. Accordingly, we will sometimes refer to the signal patterns as *fingerprints*. Fingerprinting is adopted not only for indoor environments using WLAN networks [6, 9, 10] but also for outdoor environments using WLAN and/or cellular radio access networks [11–13].

Fingerprinting consists of two phases: a training phase and a positioning phase. The training phase collects a set of RF signal patterns, used as the reference signal patterns, at reference locations within the targeted area. These reference signal patterns are stored in a database, which is called a radio map or fingerprinting database. In the positioning phase, an algorithm estimates the user position by finding the reference location at which the observed fingerprint has the best match to its reference fingerprint.

There are two common approaches to find the best match between the observed fingerprint and a reference fingerprint: the pattern recognition approach and the probabilistic approach. The pattern recognition approach finds the best match by finding the shortest Euclidian distance between the observed fingerprint and a reference fingerprint, which is based on the differences between the signal strengths of the observed and reference fingerprints. This distance is called *signal distance*. The probabilistic approach finds the best match by calculating the maximum conditional probability that, given the observed signal pattern, the user is located at a particular reference location. Further details of both approaches will be discussed in the following sections.

4.2.1 The Pattern Recognition Approach

For the pattern recognition approach, a radio signal pattern (or fingerprint) is a vector of signal identifiers, such as Cell-IDs or MAC addresses, and corresponding signal measurements (e.g., RSSI measurements). The signal measurements are typically integer values. The range of RSSI values depends on the chipset of the RF receiver, but a common range is between− 100 and 0, where values closer to 0 represent higher signal strength. In some implementations, the sign is reserved and values close to 100 represent very weak signal strength.

In the training phase of the pattern recognition approach, multiple samples of the signal strength measurements are typically collected at each reference location. The reference signal pattern is typically derived by taking the average of multiple training samples.

In the positioning phase, the signal distance between the observed signal pattern $\mathbf{y}_o$ and a reference signal pattern $\mathbf{y}_r$ at location $\mathbf{x}$ can be estimated with:

$$s(\mathbf{x}) = \left\| \mathbf{y}_o - \mathbf{y}_r \right\|$$

(4.1)

where

$$\mathbf{y}_o = \begin{bmatrix} o_1 \\ o_2 \\ \vdots \\ o_n \end{bmatrix}$$

(4.2)

and

$$\mathbf{y}_r = \begin{bmatrix} r_1 \\ r_2 \\ \vdots \\ r_n \end{bmatrix}$$

(4.3)

Here o_i ($i = 1 \ldots n$) is the observed RSSI measurement to the i-th RF transmitter, r_i is a reference RSSI measurement to the ith RF transmitter collected at reference location $\mathbf{x}$, and n is the total number of RF transmitters installed in the target area. The position estimate $\hat{\mathbf{x}}$ can then be written as:

$$\hat{\mathbf{x}} = \operatorname{argmin}_{\mathbf{x}} s(\mathbf{x})$$

(4.4)

4.2.2 The Probabilistic Approach

In the probabilistic approach, the observed signal pattern is the same as that described above. The reference signal patterns, rather than being stored directly as signal strength measurements, are stored as vectors of probability distributions of the signal measurements. The range of signal strength is typically represented with six to eight histogram bins, as shown in Figure 4.4 [9, 14, 15]. The larger the histogram value of the bin, the higher the probability that the RSSI measurement lies within the corresponding bin for a particular RF transmitter at a particular location. It has been shown that eight histogram bins are sufficient to represent a probability distribution for each RF transmitter at a particular reference location. Additional bins will not make any significant contributions to improve the positioning accuracy [9, 15].

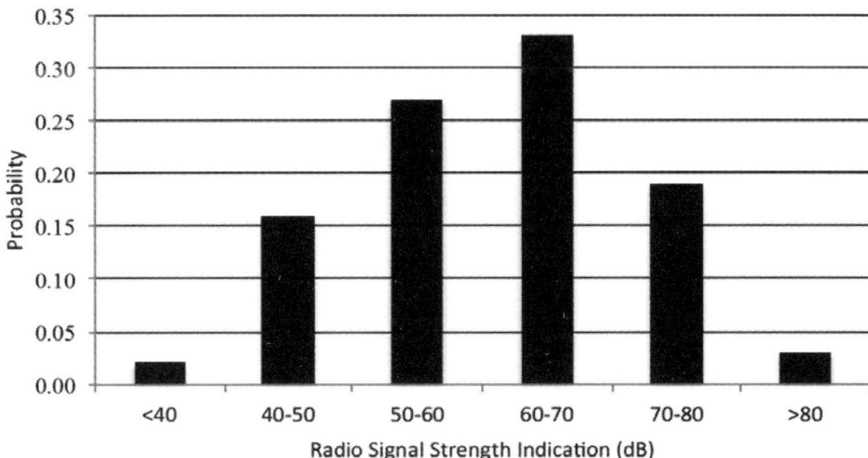

Figure 4.4 The probability distribution of a RSSI measurement.

Each probability distribution is represented by a vector of bin values; therefore, the reference signal pattern in this case becomes a vector of vectors (i.e., a matrix). In this matrix, each column is a vector representing a probability distribution for a particular RF transmitter. If there are m bins representing each probability distribution and n transmitters in a target area, the matrix will have m rows and n columns. Thus, for each reference location l the reference signal pattern is represented with an m by n signal pattern matrix $\mathbf{P}_l$ as follows:

$$\mathbf{P}_l = \begin{bmatrix} p_1^1 & p_1^2 & \cdots & p_1^n \\ p_2^1 & p_2^2 & \cdots & p_2^n \\ \vdots & \vdots & \ddots & \vdots \\ p_m^1 & p_m^2 & \cdots & p_m^n \end{bmatrix} \tag{4.5}$$

In this matrix, each value p_i^k represents a probability that the signal strength for the kth RF transmitter will fall within the range of the ith bin of the histogram when the receiver is at the location l. During the training phase, a large number of training samples are collected in order to estimate these probabilities. For the training samples collected from the kth RF transmitter falling within the ith bin, the probability p_i^k can then be estimated with:

$$p_i^k = \frac{z_i^k}{\sum\limits_{i=1}^{m} z_i^k} \tag{4.6}$$

where z_i^k is the number of training samples occurring within the measurement range of the ith bin (e.g., 40–50 dB), and m is the total number of bins.

The positioning phase is a process of finding the reference location at which the observed signal pattern occurs with the maximum probability. Instead of finding the shortest signal distance between the observed signal pattern and the reference signal patterns, the probabilistic approach uses the probabilities estimated during the training phase to find a reference fingerprint with the best match.

Given the observed signal pattern $\mathbf{y}_o = [o_1, o_2, ..., o_n]^T$, the positioning process aims to estimate the conditional probability $P(\mathbf{x}_j|\mathbf{y}_o)$ for each reference location $\mathbf{x}_j = [x_j, y_j]^T$ ($j = 1...L$). Here L is total number of reference locations in the target area. Having estimated the conditional probabilities $P(\mathbf{x}_j|\mathbf{y}_o)$ for all reference locations, the one with the maximum conditional probability is selected as the position of the mobile device [10, 16].

According to Bayes' theorem, we have:

$$P(\mathbf{x}_j \mid \mathbf{y}_o) = \frac{P(\mathbf{y}_o \mid \mathbf{x}_j) P(\mathbf{x}_j)}{P(\mathbf{y}_o)} \tag{4.7}$$

Taking the fact that $P(\mathbf{y}_o)$ and $P(\mathbf{x}_j)$ are constants, the conditional probability of the jth reference location can be rewritten as:

$$P(\mathbf{x}_j \mid \mathbf{y}_o) = c \cdot P(\mathbf{y}_o \mid \mathbf{x}_j) = c \cdot \prod_{k=1}^{n} p^k, \tag{4.8}$$

where c is a constant. Here p^k ($k = 1...n$) is the probability of the signal measurement o_k from transmitter k observed at reference location j. It can be selected from the kth column of reference signal pattern matrix $\mathbf{P}_j$. For example, if $o_k = 55$ dB, it belongs to the third bin as shown in Figure 4.4, then p^k will be selected from the third row of the kth column of $\mathbf{P}_j$.

Having estimated the conditional probabilities for all reference locations using (4.8), the estimated position $\hat{\mathbf{x}}$ of the mobile device will be the one that has the maximum probability, that is:

$$\hat{\mathbf{x}} = \arg\max_j P(\mathbf{x}_j \mid \mathbf{y}_o) \tag{4.9}$$

4.3 POSITIONING BASED ON RANGE AND RANGE DIFFERENCE

Range and range difference are common observables adopted for mobile positioning solutions in wireless radio access networks [17, 18]. A range measurement can be obtained either by measuring the signal travel time between the transmitter and the receiver and multiplying this time by the speed of light, or by converting the received signal strength to a distance using a path loss model, [e.g., the model given in (2.12)]. A range difference is the difference between two range observables. By taking the difference, the clock error of the mobile device can be removed because it is common to both range measurements.

It is common to adopt a planar approach for mobile device positioning when using RF measurements because indoor environments are typically planar (i.e., floors). Therefore, the algorithms described in this section will be given for the two-dimensional case. However, they can be easily extended to three dimensions if needed.

4.3.1 Positioning Based on Range

Using range measurements as input, the positioning solution can be determined with (3.7) or (3.14), if the clocks of all base stations or access points are synchronized (e.g., using precise clocks or GPS receivers). The positioning solution is very similar to that of the GNSS solution. The only difference is that the coordinates of the RF transmitters are fixed and can be obtained directly from a database, while that for the GPS satellites are typically calculated from the parameters transmitted in the satellite ephemerides. When taking a planar approach, the coordinates of the mobile device will be a two-dimensional vector, instead of a three-dimensional vector as shown in (3.7) and (3.14). Similar to the clock error of the GNSS receiver, the clock error of the mobile device can be determined as a parameter in the solution as well.

With range measurements obtained from a path loss model based on the received RF signal strengths, the situation is simpler because there is no clock error involved in this case. Taking a planar approach, the position of the mobile device can be located with the following mathematical model:

$$
\begin{bmatrix} d_1 \\ d_2 \\ \vdots \\ d_n \end{bmatrix} = \begin{bmatrix} \sqrt{(x_1 - x)^2 + (y_1 - y)^2} \\ \sqrt{(x_2 - x)^2 + (y_2 - y)^2} \\ \vdots \\ \sqrt{(x_n - x)^2 + (y_n - y)^2} \end{bmatrix}
\tag{4.10}
$$

where d_i is the distance between the RF transmitter and the mobile device (the receiver), (x,y) are the planar coordinates of the mobile device, and (x_i,y_i) are the planar coordinates of the ith transmitter. Theoretically, the distance d_i needs to be projected to the floor plan before applying it for position estimation. The effects of this projection, also known as a slope correction, are not significant in most cases because the ranging error is much larger than the slope correction. Taking the square of both sides of (4.10), we have:

$$\begin{bmatrix} d_1^2 \\ d_2^2 \\ \vdots \\ d_n^2 \end{bmatrix} = \begin{bmatrix} x^2+y^2+x_1^2+y_1^2-2(x_1x+y_1y) \\ x^2+y^2+x_2^2+y_2^2-2(x_2x+y_2y) \\ \vdots \\ x^2+y^2+x_n^2+y_n^2-2(x_nx+y_ny) \end{bmatrix} \quad (4.11)$$

Taking the difference of the ith equation and the first equation in (4.11), the second order terms x^2 and y^2 can then be removed. As a result, we obtain the following system of equations, which are linear in terms of x and y:

$$\begin{bmatrix} d_2^2-d_1^2 \\ d_3^2-d_1^2 \\ \vdots \\ d_n^2-d_1^2 \end{bmatrix} = \begin{bmatrix} x_2^2+y_2^2-x_1^2-y_1^2+2(x_1-x_2)x+2(y_1-y_2)y \\ x_3^2+y_3^2-x_1^2-y_1^2+2(x_1-x_3)x+2(y_1-y_3)y \\ \vdots \\ x_n^2+y_n^2-x_1^2-y_1^2+2(x_1-x_n)x+2(y_1-y_n)y \end{bmatrix} \quad (4.12)$$

We can rearrange (4.12) as follows:

$$\begin{bmatrix} x_1-x_2 & y_1-y_2 \\ x_1-x_3 & y_1-y_3 \\ \vdots & \vdots \\ x_1-x_n & y_1-y_n \end{bmatrix} \begin{bmatrix} x \\ y \end{bmatrix} = \frac{1}{2} \begin{bmatrix} x_1^2+y_1^2-x_2^2-y_2^2+d_2^2-d_1^2 \\ x_1^2+y_1^2-x_3^2-y_3^2+d_3^2-d_1^2 \\ \vdots \\ x_1^2+y_1^2-x_n^2-y_n^2+d_n^2-d_1^2 \end{bmatrix} \quad (4.13)$$

which is more compactly expressed in matrix form as follows:

$$\mathbf{Hx} = \mathbf{b} \quad (4.14)$$

where $\mathbf{H}$ is the design matrix,

$$\mathbf{H} = \begin{bmatrix} x_1 - x_2 & y_1 - y_2 \\ x_1 - x_3 & y_1 - y_3 \\ \vdots & \vdots \\ x_1 - x_n & y_1 - y_n \end{bmatrix}$$

$\mathbf{x}$ is the position parameter vector,

$$\mathbf{x} = \begin{bmatrix} x \\ y \end{bmatrix}$$

and $\mathbf{b}$ is a vector that can be estimated with the coordinates of the RF transmitters and the range measurements as follows:

$$\mathbf{b} = \frac{1}{2} \begin{bmatrix} x_1^2 + y_1^2 - x_2^2 - y_2^2 + d_2^2 - d_1^2 \\ x_1^2 + y_1^2 - x_3^2 - y_3^2 + d_3^2 - d_1^2 \\ \vdots \\ x_1^2 + y_1^2 - x_n^2 - y_n^2 + d_n^2 - d_1^2 \end{bmatrix}$$

If there are three range measurements (i.e., $n = 3$), the solution of (4.14) can then be written as:

$$\mathbf{x} = \mathbf{H}^{-1} \mathbf{b} \tag{4.15}$$

When $n > 3$, a least squares approach can be used as follows:

$$\mathbf{x} = (\mathbf{H}^T \mathbf{H})^{-1} \mathbf{H}^T \mathbf{b} \tag{4.16}$$

The covariance matrix of the estimated parameters can be estimated with:

$$\mathbf{D}_{\mathbf{xx}} = \sigma_0^2 (\mathbf{H}^T \mathbf{H})^{-1}$$

where σ_0^2 is the variance of the range observations.

By knowing the signal strength measurements to three access points or base stations, the planar location of the mobile device can be located. There are, however, large errors in the distances derived from signal strengths using the path loss model. One reason is because the effective range of the path loss model is limited. Outside of the effective range, the signal strength is not very sensitive to changes in distance (i.e., a small error in a signal strength measurement outside of the effective range can cause a large error in the distance obtained using the model).

4.3.2 Positioning Based on Range Difference

In positioning solutions based on cellular networks, range difference (as opposed to range) is often used as the positioning observable. The main advantage to using range difference is that it does not require synchronization of the clocks of the RF transmitters and the receiver (nor solving of the clock offsets). The difference of the clock errors between two transmitters (base stations or access points) can be determined by measuring the ranges measured at known locations as follows:

$$\Delta t_{21} = (r_2 - r_1) - (d_2 - d_1) \tag{4.17}$$

where Δt_{21} is difference of clock errors between the transmitters 1 and 2, r_i ($i = 1$ or 2) is the measured range to the ith transmitter and d_i ($i = 1$ or 2) is the geometric distance between the mobile device and the ith transmitter. As the mobile device is located at a known location for this purpose, the coordinates of the transmitters are known. The geometric distance d_i can then be calculated using the known coordinates. The clock error of the mobile device has been canceled by taking the difference of the ranging measurements r_1 and r_2. The differences of the clock errors determined with (4.17) are broadcast to other mobile users so that they can apply the corrections to the observed range differences before using them for positioning.

By using the Taylor's series expansion to linearize the observation equations, the following mathematical model can be derived for positioning using the range difference:

$$\begin{aligned}
\Delta \mathbf{d} &= f(\mathbf{x}) \\
&= f(\mathbf{x}_0) + \mathbf{H}\mathbf{dx}
\end{aligned} \tag{4.18}$$

where $\Delta \mathbf{d}$ is a vector of the range difference observations,

$$\Delta \mathbf{d} = \begin{bmatrix} d_1 - d_i \\ d_2 - d_i \\ \vdots \\ d_n - d_i \end{bmatrix}$$

and $\mathbf{dx}$ is the coordinate parameters to be estimated

$$\mathbf{dx} = \mathbf{x} - \mathbf{x}_0$$

or in more detail

$$\mathbf{dx} = \begin{bmatrix} dx \\ dy \end{bmatrix}, \quad \mathbf{x} = \begin{bmatrix} x \\ y \end{bmatrix}, \quad \mathbf{x_0} = \begin{bmatrix} x_0 \\ y_0 \end{bmatrix}$$

Here $\mathbf{x_0} = (x_0, y_0)$ are the approximate coordinates of the mobile device; $\mathbf{x} = (x, y)$ are its actual coordinates. The approximate range differences at the initial coordinates $\mathbf{x_0}$ can be expressed as:

$$f(\mathbf{x_0}) = \begin{bmatrix} \sqrt{(x_1 - x_0)^2 + (y_1 - y_0)^2} - \sqrt{(x_i - x_0)^2 + (y_i - y_0)^2} \\ \sqrt{(x_2 - x_0)^2 + (y_2 - y_0)^2} - \sqrt{(x_i - x_0)^2 + (y_i - y_0)^2} \\ \vdots \\ \sqrt{(x_n - x_0)^2 + (y_n - y_0)^2} - \sqrt{(x_i - x_0)^2 + (y_i - y_0)^2} \end{bmatrix}$$

where (x_j, y_j) $(j = 1...n)$ are the coordinates of the jth RF transmitter. The design matrix of the observation equation $\mathbf{H}$ can be written as:

$$\mathbf{H} = \frac{\partial f}{\partial \mathbf{x}}\bigg|_{\mathbf{x}=\mathbf{x_0}}$$

By taking the derivative of the function $f(\mathbf{x})$ and evaluating it at the initial coordinates $\mathbf{x_0}$, the matrix $\mathbf{H}$ can be obtained as follows:

$$\mathbf{H} = \begin{bmatrix} \dfrac{d_i^0(x_0 - x_1) - d_1^0(x_0 - x_i)}{d_1^0 d_i^0} & \dfrac{d_i^0(y_0 - y_1) - d_1^0(y_0 - y_i)}{d_1^0 d_i^0} \\[2mm] \dfrac{d_i^0(x_0 - x_2) - d_2^0(x_0 - x_i)}{d_2^0 d_i^0} & \dfrac{d_i^0(y_0 - y_2) - d_2^0(y_0 - y_i)}{d_2^0 d_i^0} \\[2mm] \vdots & \vdots \\[2mm] \dfrac{d_i^0(x_0 - x_n) - d_n^0(x_0 - x_i)}{d_n^0 d_i^0} & \dfrac{d_i^0(y_0 - y_n) - d_n^0(y_0 - y_i)}{d_n^0 d_i^0} \end{bmatrix}$$

where d_j^0 $(j = 1...n)$ is the approximate range to the jth RF transmitter. The least squares solution of (4.18) can be obtained by:

$$\mathbf{dx} = \left(\mathbf{H}^T\mathbf{H}\right)^{-1}\mathbf{H}^T\left(\Delta\mathbf{d} - \Delta\mathbf{d}_0\right)$$

$$\mathbf{x} = \mathbf{x}_0 + \mathbf{dx}$$

(4.19)

The covariance matrix of the estimated parameters can be estimated with:

$$\mathbf{D_{xx}} = \sigma_0^2(\mathbf{H^T H})^{-1}$$

where σ_0^2 is the variance of the range difference observations.

Depending on the closeness of the estimates coordinates $\mathbf{x}_0$ to the actual coordinates, a few iterations may need to converge to a final solution.

4.4 POSITIONING BASED ON DoA

In addition to range and range difference, DoA can also be used for positioning as was shown in Figure 2.9 [7, 8, 19]. DoA can be measured with an antenna array or a directional antenna. Using a planar approach, a DoA observable defines a straight line in the local geodetic coordinate system, while two straight lines can intersect at most to a single point (i.e., if they are not parallel). In terms of the transmitter and receiver coordinates, the DoA observable can be written as:

$$\begin{aligned}
\alpha_i &= f(x,y) \\[4pt]
&= \tan\frac{y - y_i}{x - x_i} \\[4pt]
&= \tan\frac{y_0 - y_i}{x_0 - x_i} + \left.\frac{\partial f}{\partial x}\right|_{x=x_0,y=y_0} dx + \left.\frac{\partial f}{\partial y}\right|_{x=x_0,y=y_0} dy \\[4pt]
&= a_i + b_i dx + c_i dy
\end{aligned}$$

(4.20)

where

$$a_i = \tan\frac{y_0 - y_i}{x_0 - x_i}$$

$$b_i = \left.\frac{\partial f}{\partial x}\right|_{x=x_0,y=y_0} = -\frac{y_0 - y_i}{(x_0 - x_i)^2}\sec^2\left(\frac{y_0 - y_i}{x_0 - x_i}\right)$$

$$c = \left.\frac{\partial f}{\partial y}\right|_{x=x_0,y=y_0} = \frac{1}{x_0 - x_i}\sec^2\left(\frac{y_0 - y_i}{x_0 - x_i}\right)$$

When there are more than two DoA measurements available, the following system of linear equations can be used to estimate the planar coordinates of the mobile device:

$$\begin{bmatrix} \alpha_1 \\ \alpha_2 \\ \vdots \\ \alpha_n \end{bmatrix} - \begin{bmatrix} a_1 \\ a_2 \\ \vdots \\ a_n \end{bmatrix} = \begin{bmatrix} b_1 & c_1 \\ b_2 & c_2 \\ \vdots & \vdots \\ b_n & c_n \end{bmatrix} \begin{bmatrix} dx \\ dy \end{bmatrix} \tag{4.21}$$

or in matrix form

$$\mathbf{l} = \mathbf{H}\mathbf{dx} \tag{4.22}$$

where

$$\mathbf{l} = \begin{bmatrix} \alpha_1 \\ \alpha_2 \\ \vdots \\ \alpha_n \end{bmatrix} - \begin{bmatrix} a_1 \\ a_2 \\ \vdots \\ a_n \end{bmatrix}$$

$$\mathbf{H} = \begin{bmatrix} b_1 & c_1 \\ b_2 & c_2 \\ \vdots & \vdots \\ b_n & c_n \end{bmatrix}$$

$$\mathbf{dx} = \begin{bmatrix} dx \\ dy \end{bmatrix}$$

Similar to (4.19), (4.22) can be used to obtain a least squares solution of the coordinates of the mobile device.

4.5 SUMMARY

This chapter discussed various mobile positioning solutions using observables from RF signals originally intended only for wireless communications. Different positioning algorithms were presented, including those based on RF signal coverage area, RF signal patterns, range, ranging difference, and DoA. These positioning solutions can be applied indoors and outdoors for different wireless networks, such as wireless radio access networks (i.e., cellular networks), WLAN, or other wireless sensor networks. The algorithms described form the fundamentals to the presented positioning methods. Therefore, they can be used as a starting point for developing a more comprehensive positioning solution for mobile devices.

References

[1] Trevisani, E., and A. Vitaletii, "Cell-ID location, technique, limits and benefits: an experimental study," *Proceedings of the Sixth IEEE Workshop on Mobile Computing Systems and Applications WMCSA2004, 51-60,* 2004.

[2] Wirola, L., *Study on Location Technology Standard Evolution in Wireless Networks* (PhD thesis, Tampere University of Technology, Tampere, Finland). Retrieved from http://dspace.cc.tut.fi/dpub/handle/123456789/6509, 2010.

[3] Chen, R., *Ubiquitous Positioning and Mobile Location-Based Services in Smart Phones*, doi: 10.4018/978-1-4666-1827-5. Pennsylvania, US. IGI-Global, 2012.

[4] Llombart M., M. Ciurana, and F. Barcelo-Arroyo, "On the scalability of a novel WLAN positioning system based on time of arrival measurement," *Proceedings of IEEE 5th Workshop of Positioning, Navigation and Communication*, 15-21, 2008.

[5] Liu, J., et al., "Accelerometer assisted robust wireless signal positioning based on a hidden Markov model," *Proceedings of the IEEE/ION Position Location and Navigation Symposium 2010,* 488-497, 2010.

[6] Bahl, P., and V. N. Padmanabhan, "Radar: An in-building RF-based user location and tracking system," *Proceedings of the IEEE INFOCOM 2000 Conference on Computer Communications:* Vol. 2, 775-785, 2000.

[7] Klukas, R., and M. Fattousche, "Line-of-sight angle of arrival estimation in the outdoor multipath environment," *IEEE Transactions on Vehicular Technology,* Vol. 47, No. 1, 342-351,1998.

[8] Sakagami, S., et al., "Vehicle Position Estimation by Multibeam Antennas in Multipath Environments," *IEEE Transactions on Vehicular Technology.* 41(1), 63-68, 1992.

[9] Chen, R., L. Pei, and H. Lelppäkoski, "WLAN and Bluetooth positioning in smart phones," In *Ubiquitous Positioning and Mobile Location-Based Services in Smart Phones*, R. Chen (ed), DOI: 10.4018/978-1-4666-1827-5.ch003. Pennsylvania, US. IGI-Global. 2012.

[10] Roos, T., et al., "A probabilistic approach to WLAN user location estimation," *International Journal of Wireless Information Networks,* 9(3), 155-164, 2002.

[11] Zandbergen, P. A., "Accuracy of iPhone locations: a comparison of assisted GPS, WLAN and cellular positioning," *Transactions in GIS, 13(*s1*):* 5–26, 2009.

[12] Ahonen, S., and P. Eskeinen, "Mobile terminal location for UMTS," *IEEE Aerospace and Electronic Systems Magazine.* 18(2), 23-27, 2003.

[13] Lakmali, B.D.S. and D. Dias, "Database correlation for GSM location in outdoor & indoor environments," *Proceedings of the 4th International Conference on Information and Automation for Sustainability, ICIAFS 2008,* 42-47, 2000.

[14] Pei, L., et. al., "Using inquiry-based Bluetooth RSSI probability distributions for indoor positioning," *Journal of Global Positioning Systems,* 9(2), 122-130, 2010.

[15] Leppäkoski, H., S. Tikkinen, and J. Takala, "Optimizing radio map for WLAN fingerprinting," *Proceedings of Ubiquitous Positioning Indoor Navigation and Location Based Service (UPINLBS),* 1-8, 2010.

[16] Youssef, M., A. Agrawala, and A. U. Shankar, "WLAN location determination via clustering and probability distributions," *Proceedings of the First IEEE International Conference on Pervasive Computing and Communications,* 143-150, 2003.

[17] Zhao, Y., "Standardization of mobile phone positioning for 3G systems," *IEEE Communication Magazine,* 108-116, July 2002.

[18] Bull, J. F., "Advantages and Disadvantages of the Two Basic Approaches for E-911," *IEEE Vehicular Technology Magazine,* 45-53, December 2009.

[19] Belloni, F., et al., "Angle-based indoor positioning system for open indoor environments," *Proceedings of the 6th Workshop on Positioning, Navigation and Communication,* 261-265, 2009.

Chapter 5

Hybrid Positioning in Mobile Devices

In previous chapters, we have discussed positioning solutions based on GNSS technology and wireless networks. Some of these positioning solutions work well outdoors (e.g., the GNSS solutions), while others work well indoors (e.g., the WLAN solutions). To achieve a ubiquitous positioning solution (i.e., indoors and outdoors) with satisfactory positioning accuracy, a hybrid approach is needed.

Hybrid positioning solutions integrate different types of positioning sensors, various RF signals, and pedestrian gait parameters. Compared to positioning solutions based on a single positioning technology (e.g., GNSS), a hybrid positioning solution improves the availability of a position fix significantly, especially for indoor environments.

Hybrid positioning solutions in mobile devices typically utilize the built-in GNSS receiver, various motion and orientation sensors, and RF signals. The built-in motion and orientation sensors include accelerometers, gyroscopes, digital compasses, and cameras, while the RF signals include signals from various short-range RF networks, especially WLAN networks because of their widespread deployment worldwide.

A hybrid positioning solution usually consists of one or more absolute positioning technologies and one or more relative positioning technologies. GNSS is the most common absolute positioning technology for outdoor environments, whereas WLAN-based positioning technology is the most widely used technology for indoor environments. Accelerometers and gyroscopes are the most common sensors used for generating a relative positioning solution. These sensors have very high sampling rates (more than 100 Hz for the sensors in mobile devices), but the measurements suffer from errors that cause the positioning solution to drift after a short while. Therefore, they are very beneficial for producing a positioning solution with high temporal resolution during the gap between two absolute positions, but they cannot be used indefinitely without calibrations from another source. In addition to sensors and RF signals, pedestrian gait parameters, such as step length and step frequency, are also common parameters adopted for relative positioning using the PDR approach. PDR will be presented in more detail in Section 5.2.

Typical approaches for hybrid positioning solutions can be divided into three categories: 1) integration of sensor measurements and pedestrian gait parameters, 2) integration sensor measurements and RF signals, and 3) integration of camera, sensor, GNSS receiver, and RF signals. The first approach is known as PDR; the second approach is called multisensor, multisignal positioning; and the third approach is called visual-based or visual-aided positioning.

We start the chapter with a simple demonstration of the availability of the sensor measurements in mobile devices. Section 5.2 then presents the PDR solution detail, followed by Section 5.3, which discusses the multisensor, multisignal approach. Section 5.4 describes visual-based and visual-aided positioning, and the chapter concludes with a short summary in Section 5.5.

5.1 MEASUREMENTS OF THE BUILT-IN SENSORS IN MOBILE DEVICES

In addition to a GNSS receiver, motion and orientation sensors are standard hardware components in mobile devices. These include accelerometers, digital compasses, and gyroscopes. The raw measurements of these sensors can be accessed easily through application programing interfaces (APIs) in all mobile platforms, including Android, iOS, and Windows Phone. Figure 5.1 shows a simple application implemented in the Android mobile platform. The software reads the measurements from the built-in sensors, including the GNSS positions and raw data from the accelerometers, gyroscopes, and magnetometer (i.e., digital compass). In addition to the sensor measurements, the RSSIs from surrounding WLAN access points are also displayed. The RSSI values are scaled to a range from 1 to 10. All these measurements are fundamental inputs for developing a hybrid positioning solution.

The sample rates for the built-in inertial sensors can be as high as 180 Hz (e.g., the Samsung Galaxy 4S smartphone). Other measurements, such as the GNSS and WLAN receivers, have much lower sample rates (in terms of output position or WLAN scan). For GNSS, the typical sample rate is about 1 Hz, while for RSSI measurements it is typically less than 1 Hz. Some of the more advanced low-cost GNSS receivers (e.g., the uBlox 7) can output positions at a rate of 10 Hz. As mobile technology is now developing at a rapid pace, the sensor and GNSS measurements will continue to be enhanced in terms of quality, sample rate, and efficiency (e.g., low power consumption).

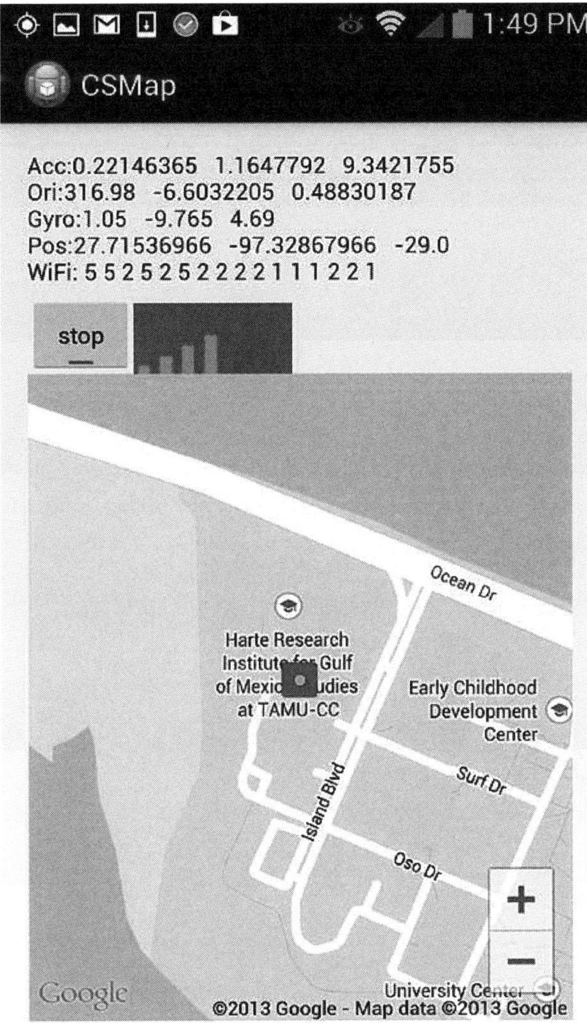

Figure 5.1 Measurements of the mobile device's built-in sensors and RSSIs from the surrounding WLAN access points. The WLAN RSSI values are scaled to a range from 1 to 10.

5.2 PDR

The PDR solution adapts the concept of "human as a sensor" for positioning. It utilizes sensor measurements to derive human gait parameters, such as step length and step frequency. It then applies the human gait parameters to propagate the user's position from a known position. The PDR solution consists of the three major steps: step detection, heading determination, and position propagation.

5.2.1 The Generic Position Propagation Model

The generic position propagation model used to generate a PDR solution is described as follows:

$$N_{k+1} = N_k + s_k \cdot \Delta t \cdot \cos \alpha_k$$
$$E_{k+1} = E_k + s_k \cdot \Delta t \cdot \sin \alpha_k \tag{5.1}$$

where N_k and E_k are the north and east coordinate components at epoch k, s_k is the walking speed of the pedestrian at epoch k, Δt is the time interval between epoch k and epoch $k+1$, and α_k is the walking heading (azimuth) of the pedestrian at epoch k. The epochs are often defined stepwise, such that each new step marks a new epoch. In this case, the term $s_k \cdot \Delta t$ can be replaced with a step length parameter SL_k. Equation (5.1) then becomes:

$$N_{k+1} = N_k + SL_k \cdot \cos \alpha_k$$
$$E_{k+1} = E_k + SL_k \cdot \sin \alpha_k \tag{5.2}$$

The position can then be propagated based on the number of steps measured by the sensors, the step length for each step, and the heading measurement for each step.

In order to propagate the position, an initial position of the user needs to be defined. This can be obtained from an absolute positioning technology, such as GNSS for outdoor environments or WLAN-based positioning for indoor environments. An initial position can also be obtained via user input. This may be necessary in cases where there is no absolute positioning technology available or the accuracy of the absolute positioning solution is too low (e.g., the Cell-ID solution from cellular networks). Any error in the initial position will remain as an error (i.e., offset) in the successive propagated positions until the system is recalibrated with a new absolute position. As a result, the user trajectory will be shifted by the offset independent of the accuracy of the PDR solutions.

As seen from (5.2), speed and heading are also required to propagate the pedestrian position. These two parameters are used to estimate the change in coordinates from epoch k to epoch $k+1$. The motion and orientation sensors integrated into mobile devices are low-cost sensors, compared to those traditionally used in navigation systems. The measurement noise is typically very large. Furthermore, the magnetometer is very sensitive to ferrous materials and electrical devices in the sensor's surroundings.

As explained in Section 2.2.3, speed can theoretically be measured by integrating accelerometer measurements. Due to the large noise component, however, it is difficult to obtain a reliable speed measurement using this approach. Instead, step length and step frequency are typically used to derive a more reliable walking speed.

Similarly, Section 2.2.4 describes the problems inherent in using either a digital compass or gyroscopes in isolation to obtain heading measurements. Fortunately, the errors from a digital compass and gyroscopes are of different types. A digital compass suffers from large episodic error fluctuations, whereas a gyroscope suffers from random white noise. Thus, fusion of the gyroscope and magnetometer measurements is the most common approach for determining the heading. Additionally, building layouts are often used to constrain the heading solution to more likely values, in order to obtain a more accurate user trajectory.

5.2.2 Step Detection

Step detection is the procedure used to determine the number of steps walked and the length of each step. The cyclic pattern of the total acceleration [see (2.17)] is the most common observable that is used for step detection. The total acceleration has the desirable property that it is independent of the orientation of the mobile device. While a pedestrian walks, the total acceleration of the mobile device forms a cyclic pattern as was shown in Figure 2.6. Each cycle of the waveform represents a single step of the pedestrian. Therefore, the total number of steps walked by the pedestrian can be estimated by counting the total number of waveform cycles. On the other hand, the frequency of the waveform represents the step frequency. Step frequency is highly correlated with step length, so it can also be used to estimate the step length, a procedure that will be described below.

There are various approaches to detect the steps. These approaches include peak detection [1], zero-crossing [2], autocorrelation [3], fast Fourier transformation [4], and direct step measurement using sensors mounted on the shoes [5].

In addition to detecting steps, the step length must be estimated. There are various approaches to model step length. These approaches can be divided into four major categories: the constant model approach, the linear model approach, the nonlinear model approach, and the artificial neural network approach. Different models consist of different parameters; therefore, a model-training phase is typically needed for estimating the model parameters before applying the model for estimating step length.

The simplest approach is the constant model approach that makes an assumption of a constant step length, for example, 0.7m. This value can be measured for an individual or set of individuals during a training process, or a "global" step length value can be adopted based on existing knowledge. The main factors affecting step length include the height of the pedestrian and the walking pace (i.e., step frequency). Therefore, the best accuracy will be achieved when the step length value is customized for specific individuals based on empirical data. Because the constant model approach does not take into account step frequency, it may perform worse than some of the other approaches described below.

In the linear model approach, the step length can be simply defined as a linear function of the height of the pedestrian:

$$SL = a \cdot h \tag{5.3}$$

where SL is the step length, a is the model parameter, and h is the height of the pedestrian. This model is relatively simple, and it can provide a better estimate than when a global value is used in the constant model approach.

The linear model approach can be expanded by considering step frequency [6], as follows:

$$SL = (0.7 + a(h - 1.75) + \frac{b(SF - 1.79)h}{1.75})c \tag{5.4}$$

where SF is the step frequency, a and b are model parameters, and c is a calibration factor that is used to calibrate the model during the positioning phase. This model incorporates the fact that the step length will increase proportionate to the step frequency. Research has shown that this model can achieve an accuracy level within 5% of the traveled distance. An example of the trained model parameters are $a = 0.371$, $b = 0.227$, when the model was trained with 11 people, including five women and six men [6]. The calibration factor c is set initially to one, but if absolute positions are obtained during the positioning phase, it can be adjusted in order to calibrate the model based on the known distance traveled.

Renaudin et al. [7] proposed a similar model as follows:

$$SL = h(a \cdot SF + b) + c \tag{5.5}$$

where a, b, and c are model parameters. This model performs similarly to the model given in (5.4).

In the nonlinear approach, Fang et al. [1] used the following nonlinear function to estimate step length:

$$SL = k \cdot \sqrt[4]{a_{\max} - a_{\min}} \tag{5.6}$$

where k is a model parameter, while $a_{\max}$ and $a_{\min}$ are the maximum and minimum acceleration values at a single step (obtained from the accelerometers). The model parameter k ranges from 0.5 to 0.57 meters depending on the length of the pedestrian's legs. The performance of this model is to within 8% of the traveling distance [1]. Furthermore, this model has only one free parameter, making it easier to perform training, compared to the linear models shown above.

All models presented so far (except the constant model) require mathematical definitions of the models. The definitions of the models are empirical that are not based on theory. Researchers have been trying to find a solution to avoid the mathematical definition by adapting the artificial neural network (ANN) approach [8]. The most significant advantage of the ANN approach is that it is not required to find a mathematical function to map the relationship between the step length and the other variables such as step frequency or, maximum and minimum accelerations. Instead of training the model parameters, the ANN approach requires training a neural network.

Finally, to give an example of estimating step length, a short walking experiment was carried out in a corridor of a typical office building. The height of the person walking was 1.73m. The accelerations were recorded with a mobile device with a data sample rate of 120 Hz. The raw data was smoothed using a moving average approach, where the window size was 0.7 seconds. Figure 5.2 shows the cyclic pattern of the total acceleration during a 5-second period of walking. As we can see from the figure, the step frequency is about 2 Hz. Plugging the step frequency (SF) and user height (h) into (5.4), we obtain a step length of 0.74m for this short walking window of 5 seconds.

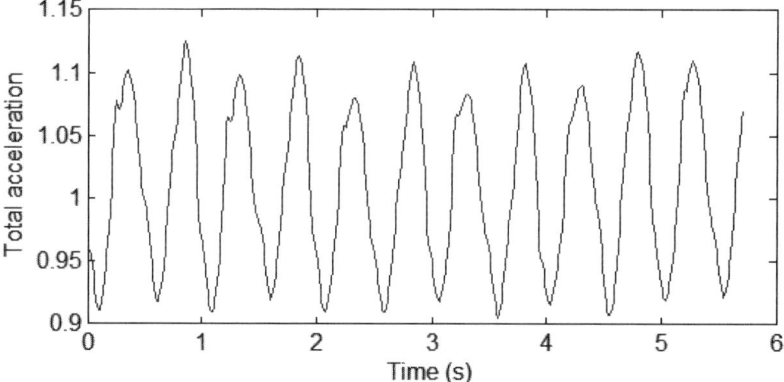

Figure 5.2 The cyclic pattern of the total acceleration during a pedestrian walking process. The data was sampled with an iPhone 4S at a rate of 120 Hz. The raw data was smoothed using a sliding-window approach, where the window size is 0.7 seconds.

5.2.3 Heading Determination

Orientation sensors, such as magnetometers and gyroscopes, are the sensors most commonly used for heading determination. As described in Section 2.2.4, a magnetometer can output absolute orientation, while gyroscopes deliver angular rates. While the magnetometer does not suffer from drift, it is very sensitive to magnetic perturbations, which can affect heading determination. Such perturbations are particularly severe in indoor environments, where ferrous materials and electrical devices such as elevators, steel staircases, computers, copy machines, and printers disturb the local magnetic field. Because the most common environment for PDR solutions is indoors, heading determination using a magnetometer in isolation is a challenging task. Figure 5.3 shows the magnetometer readings along a straight corridor in a typical office building. The absolute heading error can be as large as tens of degrees.

On the other hand, gyroscopes are not affected by magnetic perturbations. When used to determine heading, however, the integration of gyroscope measurements leads to *drift*. Specifically the error in absolute orientation increases linearly over time. Over short periods of time, however, gyroscopes can provide fairly high-quality orientation measurements, and these measurements can be used to discern whether fluctuations in the magnetometer output are due to orientation changes or rather from magnetic disturbances. Therefore, the magnetometer and gyroscope are complementary sensors.

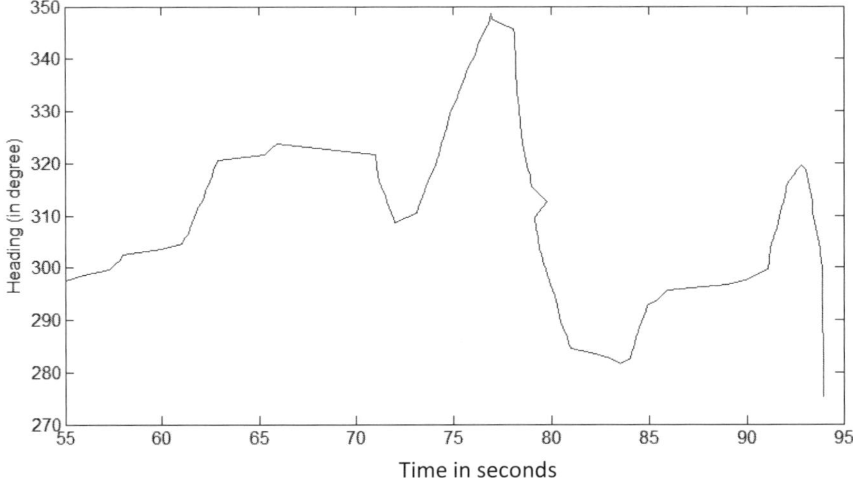

Figure 5.3 Heading readings obtained from a digital compass in a smartphone for a straight corridor.

In order to fuse magnetometer and gyroscope data, measurements from both sensors are first smoothed using a moving average approach. Data smoothing can reduce the impact of random errors. In case there is no significant magnetic perturbation in the positioning environment, the smoothed magnetometer and gyroscope measurements will show a strong correlation. In this case, it is safe to apply the smoothed heading measurements from the magnetometer for heading determination, as well as calibration of the gyroscope.

In case there is a significant difference between the two datasets over a short period, it is an indication that magnetic perturbations exist in the sensor's surroundings. During these periods, the heading should be updated using only the smoothed gyroscope measurements. This approach cannot solve the problem of magnetic perturbations completely; it can only enhance the solution of heading determination during short periods, such as a pedestrian briefly passing a "hot spot" of magnetic perturbation.

The combination of the gyroscope and magnetometer measurements is a self-contained approach for heading determination using the built-in sensors. It can be further enhanced with the use of building layout information. The orientation of straight corridors can be obtained from building layout drawings. In order to apply the building layout constraint to the heading measurements, we need to identify the corridor through which the pedestrian is currently walking. This can be done by, for example, using the WLAN positioning solution or even by mapping the significant magnetic perturbations as landmarks in a building's environment [9]. While magnetic perturbation is viewed as an error source with respect to a magnetometer measurement, it can be a "signal" for identifying the location of a significant landmark as well.[14] The building layout constraint works well for environments with narrow corridors where the pedestrian's heading can be constrained to only two directions: direction A and direction $A+180°$. It does not work, however, in open areas where the pedestrian can walk in all directions.

5.2.4 Position Propagation

Positions along the pedestrian's traveling trajectory can be propagated with (5.1) or (5.2). We need only two variables to propagate the position from epoch k to epoch $k+1$: the translation d_k and the heading α_k at epoch k. The translation can be obtained either by:

$$d_k = s_k \cdot \Delta t$$
$$= s_k(t_{k+1} - t_k)$$

or by the step length SL_k, estimated using the techniques described in Section 5.2.2. The traveling speed s_k is the product of the step length (SL) and step frequency (SF). When (5.2) is used, the epochs should be defined stepwise, but if (5.1) is used, the time interval can be chosen more freely.

The error of the propagated position grows with the distance traveled because of the errors in the estimated step length and heading. Any systematic error in step length will stretch or shrink the shape of the trajectory, while a systematic error in heading will cause rotations of the trajectory with respect to the true one. Therefore, we need to calibrate the propagated trajectory whenever a known position is available. The calibration will 1) reinitiate the starting position of the

---

[14] In fact, using magnetic field anomalies as an observable for absolute positioning is an active research topic. See [9] for details.

PDR solution, and 2) calibrate the step length model, for example, by adjusting the calibration factor c in (5.4). The first action eliminates the cumulative error in the propagated position, while the second action will reduce the systematic error in future step length estimations, thus reducing the error growth as the reinitialized position is propagated. The absolute positions can be obtained from absolute positioning solutions (e.g., GNSS solutions, WLAN positioning solutions, or position solutions obtained from RFID and NFC proximity-based solutions).

Figure 5.4 shows the result of a PDR solution obtained from a typical office environment at Texas A&M University Corpus Christi. The true walking path was a rectangle consisting of four corridors forming a closed-loop (i.e., the starting and ending points were the same).

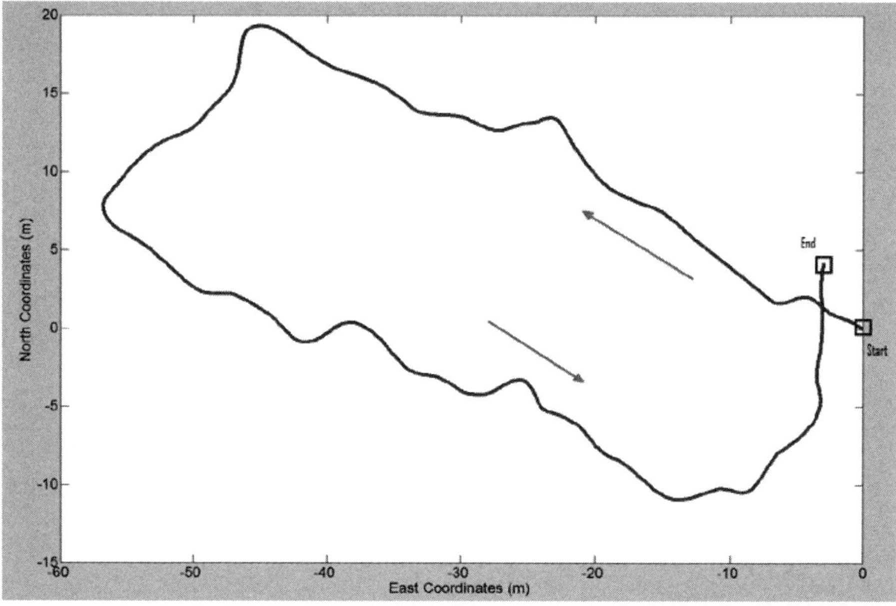

Figure 5.4 The trajectory of a PDR solution based on smartphone sensors. The true walking path is a closed loop of four straight corridors.

The PDR solution was purely based on sensor measurements from a mobile device (Samsung Galaxy S4). No building layout constraints were applied in the PDR solution. Step detection and step length estimation were based on accelerometer measurements, while headings were estimated by fusing measurements of the magnetometer and gyroscopes. The step lengths were estimated with (5.4). From this figure, the following two points become clear:

- The walking paths derived from the PDR solution are not straight lines. This is likely because of magnetic perturbations along the corridors.
- The estimated walking trajectory does not form a closed loop. This is likely caused by a combination of the scale error of the step-length model and the errors in heading determination.

The difference in position between the starting and ending points in this experiment is about 5m. Had position updates been applied (e.g., from a WLAN-based positioning solution), this error would likely have been less. Furthermore, if building layout constraints had been applied to constrain the heading result, the trajectory would have a rectangular form more closely resembling the true path. Despite these shortcomings, the sensor-only solution performs reasonably well, considering that the low-cost sensors in mobile devices were not intended for positioning purposes.

5.3 MULTISENSOR MULTISIGNAL POSITIONING

The PDR solution integrates motion and orientation sensors with pedestrian locomotion to achieve a hybrid solution. It uses absolute positioning technologies only to provide an initial solution and optionally to recalibrate the mobile position periodically. In this section we will address more advanced methods of integrating sensors with absolute positioning technologies such as GNSS and WLAN.

5.3.1 Integration Strategy

In precision navigation systems, such as those used in aircraft and in military applications, it is common to integrate GNSS-based positioning with inertial sensor measurements, a technique called GNSS/inertial navigation system (INS) integration. This approach is one of the most successful examples of integrating an absolute positioning technology with motion and orientation sensors. GNSS/INS integration is typically performed using one of two different integration approaches, known respectively as the loosely coupled and tightly coupled approaches. The loosely coupled approach integrates the data from the two systems in the coordinate domain, meaning that each system independently calculates the user position (although a solution from one system may provide an initial result for the other system). In fact, PDR can be considered a loosely coupled approach. In the tightly coupled approach, measurements from one system can be used to help correct or interpret measurements from the other system, prior to any position solution being obtained. This usually enabled a more accurate overall positioning solution.

Unfortunately, the quality of current sensors in mobile devices is not high enough to allow us to implement a tightly coupled solution. Therefore, the loosely coupled approach is more suitable for mobile devices. It offers more flexibility to handle the heterogeneous nature of the sensors and signals adopted in the solution. Figure 5.5 shows the general architecture of a loosely coupled approach for hybrid positioning, based on various sensors and RF signals in mobile devices. Even though we did not explicitly show this architecture when discussing the PDR solution above, its structure should look somewhat familiar.

In addition to the PDR solution, another way to implement the loosely coupled approach is with a Kalman filter. The Kalman filter was described in Chapter 3 with respect to GNSS-based positioning. One advantage of the Kalman filter is that the equations are the same whether the measurements come from GNSS, a WLAN-based solution, motion and orientation sensors, or any other available source. So long as an error estimate of the measurements can be obtained, measurements from any positioning system can be applied.

For simplicity, we will consider measurements from three different types of sensors in the integration algorithm described in Section 5.3.2:

- Accelerometers, which provide a speed measurement for the Kalman filter. The speed estimate can be obtained by step detection as explained in Section 5.2.2. Recall that speed is the product of the step length and step frequency, where the latter can be derived from signal of the total acceleration.
- Gyroscopes and magnetometers, which together provide heading and heading rate measurements for the Kalman filter. The steps to integrate gyroscope and magnetometer measurements to obtain heading and heading rate were described in Section 5.2.3. This is considered to be a preprocessing step in the Kalman filter algorithm.
- Absolute positioning technologies, such as GNSS and WLAN-based positioning solutions. These primarily provide a position estimate to the Kalman filter, but they can also be used to provide a speed estimate.

5.3.2 Integration Algorithm

Next we will present the integration algorithm that implements the loosely coupled approach using a Kalman filter. Since this approach is typically used in indoor environments and mobile users in these environments are mainly constrained to two-dimensional planes (i.e., floors), the algorithm will consider only two-dimensional positions. As indicated in Figure 5.5, the algorithm is usually implemented in a hybrid integration block, whose inputs are the following observables:

- Two-dimensional coordinates (N, E), which can be obtained either from a GNSS or WLAN-based positioning solution;
- Heading obtained from the fusion of magnetometer and gyroscope measurements;
- Heading rate also obtained from the fusion of magnetometer and gyroscope measurements;

 —   Speed obtained from accelerometers via step detection.

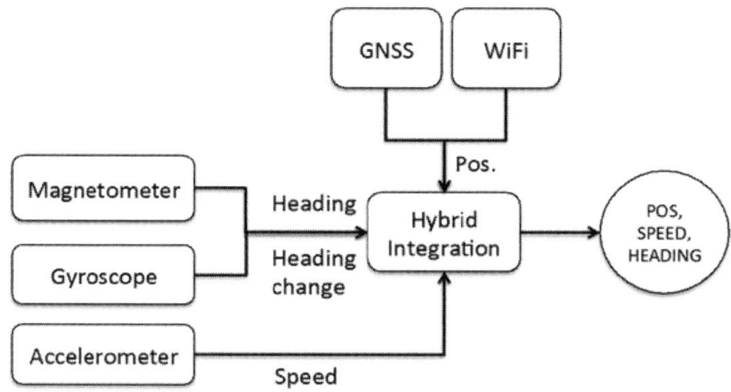

Figure 5.5 Integration scheme of the loosely coupled hybrid positioning solution.

The state vector of the Kalman filter is defined as:

$$\mathbf{x} = \begin{bmatrix} N \\ E \\ \dot{a} \\ \alpha \\ s \end{bmatrix} \tag{5.7}$$

where N and E are the north and east coordinate components, α is the heading with an origin pointing east and counterclockwise positive, $\dot{a}$ is the heading rate, and s is the speed.

The system process model, which describes how the state vector changes from one time epoch to the next, is defined as follows [10]:

$$\begin{aligned}
N_{k+1} &= N_k + s_k \cdot \Delta t \cdot \sin \alpha_k + w_1 \\
E_{k+1} &= E_k + s_k \cdot \Delta t \cdot \cos \alpha_k + w_2 \\
\dot{\alpha}_{k+1} &= \dot{\alpha}_k + w_3 \\
\alpha_{k+1} &= \alpha_k + \dot{\alpha}_k \cdot \Delta t + w_4 \\
s_{k+1} &= s_k + w_5
\end{aligned} \tag{5.8}$$

where k indexes the time epoch of the measurement, Δt is the time difference between epoch k and k + 1, and w_i are random process noise components.

The system process model can also be expressed in matrix form as:

$$\mathbf{x}_{k+1} = \mathbf{\Phi}_k \mathbf{x}_k + \mathbf{w}_k, \qquad \mathbf{w}_k \sim N(0, \mathbf{Q}_k) \tag{5.9}$$

where $\mathbf{\Phi}_k$ is the state transition matrix and $\mathbf{w}_k$ is the process noise vector, which is normally distributed with zero mean and covariance $\mathbf{Q}_k$. A model for the covariance $\mathbf{Q}_k$ will be described later in this section. To avoid linearization, the state transition matrix $\mathbf{\Phi}_k$ can be obtained from (5.8) directly as follows [10]:

$$\mathbf{\Phi}_k = \begin{bmatrix} 1 & 0 & 0 & s \cdot \cos \alpha_k \cdot \Delta t & \sin \alpha_k \cdot \Delta t \\ 0 & 1 & 0 & -s \cdot \sin \alpha_k \cdot \Delta t & \cos \alpha_k \cdot \Delta t \\ 0 & 0 & 1 & 0 & 0 \\ 0 & 0 & \Delta t & 1 & 0 \\ 0 & 0 & 0 & 0 & 1 \end{bmatrix} \tag{5.10}$$

Next, the observation model, which describes how the measurements (i.e., position, speed, heading, and heading change) relate to the true values of the state vector, is the general linear model as follows:

$$\mathbf{z}_k = \mathbf{H}\mathbf{x}_k + \mathbf{v}_k, \qquad \mathbf{v}_k \sim N(0, \mathbf{R}_k) \tag{5.11}$$

where $\mathbf{z}_k$ is the measurement vector, $\mathbf{H}$ is the design matrix, and $\mathbf{v}_k$ is the observation noise (i.e., measurement error), which is normally distributed with zero mean and covariance $\mathbf{R}_k$. The measurement vector $\mathbf{z}_k$ is defined as:

$$\mathbf{z}_k = \begin{bmatrix} N_{GNSS} \\ E_{GNSS} \\ N_{WiFi} \\ E_{WiFi} \\ \dot{\alpha} \\ \alpha \\ s_{GNSS} \\ s_{accel} \end{bmatrix} \tag{5.12}$$

Note that the measurement vector contains multiple sources for the measurements of north and east position components (GNSS and WLAN), as well as speed measurements from both GNSS and accelerometers. If additional measurement sources are available, the measurement vector can be updated accordingly. The corresponding design matrix $\mathbf{H}$ can be expressed as:

$$\mathbf{H} = \begin{bmatrix} 1 & 0 & 0 & 0 & 0 \\ 0 & 1 & 0 & 0 & 0 \\ 1 & 0 & 0 & 0 & 0 \\ 0 & 1 & 0 & 0 & 0 \\ 0 & 0 & 1 & 0 & 0 \\ 0 & 0 & 0 & 1 & 0 \\ 0 & 0 & 0 & 0 & 1 \\ 0 & 0 & 0 & 0 & 1 \end{bmatrix} \tag{5.13}$$

For simplicity, the covariance matrix of the observation noise $\mathbf{R}_k$ can be approximated with a diagonal matrix:

$$\mathbf{R}_{k+1} = \begin{bmatrix} \sigma_{N-GNSS}^2 & 0 & 0 & 0 & 0 & 0 & 0 & 0 \\ 0 & \sigma_{E-GNSS}^2 & 0 & 0 & 0 & 0 & 0 & 0 \\ 0 & 0 & \sigma_{N-WiFi}^2 & 0 & 0 & 0 & 0 & 0 \\ 0 & 0 & 0 & \sigma_{E-WiFi}^2 & 0 & 0 & 0 & 0 \\ 0 & 0 & 0 & 0 & \sigma_{\dot{\alpha}}^2 & 0 & 0 & 0 \\ 0 & 0 & 0 & 0 & 0 & \sigma_{\alpha}^2 & 0 & 0 \\ 0 & 0 & 0 & 0 & 0 & 0 & \sigma_{s-GNSS}^2 & 0 \\ 0 & 0 & 0 & 0 & 0 & 0 & 0 & \sigma_{s-Acc}^2 \end{bmatrix} \tag{5.14}$$

where the diagonal elements are the variances of the corresponding measurements, estimated based on knowledge about the sensors or systems being used. For example, GNSS receivers often output a position and speed error estimate. Heading and heading rate errors can be derived based on the sensor hardware specifications or estimated approximately based on the correlation between the magnetometer and gyroscope measurements.

Next we define a model for covariance of the process noise $\mathbf{Q}_k$ from (5.9). Following the implementation in [11],

$$\begin{aligned} \mathbf{Q}_k &= E\{\mathbf{w}_k \mathbf{w}_k^T\} \\ &= \int_{t_k}^{t_{k+1}} \Phi_k(t_k, \varsigma) \mathbf{B}_k \Phi_k^T(t_k, \varsigma) d\varsigma \end{aligned} \tag{5.15}$$

where $\mathbf{B}_k$ is a 5×5 diagonal matrix that contains predefined spectral densities of the state parameters in its diagonal elements. Below these five diagonal elements are labeled q_i, $i = 1\ldots5$, starting from the upper-left element. These parameters q_i are usually defined based on empirical testing of a particular class of sensors. Equation (5.12) is evaluated as follows [10, 11]:

$$\mathbf{Q}_k = \begin{bmatrix} Q_{11} & Q_{12} & 0 & Q_{14} & Q_{15} \\ Q_{21} & Q_{22} & 0 & Q_{24} & Q_{25} \\ 0 & 0 & Q_{33} & Q_{34} & 0 \\ Q_{41} & Q_{42} & Q_{43} & Q_{44} & 0 \\ Q_{51} & Q_{52} & 0 & 0 & Q_{55} \end{bmatrix} \tag{5.16}$$

where

$$Q_{11} = q_1 \cdot \Delta t + \frac{a^2}{3} \cdot q_4 \cdot \Delta t^3 + \frac{b^2}{3} \cdot q_5 \cdot \Delta t^3$$

$$Q_{12} = Q_{21} = \frac{a \cdot c}{3} \cdot q_4 \cdot \Delta t^3 + \frac{b \cdot d}{3} \cdot q_5 \cdot \Delta t^3$$

$$Q_{14} = Q_{41} = \frac{a}{2} \cdot q_4 \cdot \Delta t^2$$

$$Q_{15} = Q_{51} = \frac{b}{2} \cdot q_5 \cdot \Delta t^2$$

$$Q_{22} = q_2 \cdot \Delta t + \frac{c^2}{3} \cdot q_4 \cdot \Delta t^3 + \frac{d^2}{3} \cdot q_5 \cdot \Delta t^3$$

$$Q_{24} = Q_{42} = \frac{c}{2} \cdot q_4 \cdot \Delta t^2$$

$$Q_{25} = Q_{52} = \frac{d}{2} \cdot q_5 \cdot \Delta t^2$$

$$Q_{33} = q_3 \cdot \Delta t$$

$$Q_{34} = Q_{43} = \frac{q_3}{2} \Delta t^2$$

$$Q_{44} = q_4 \cdot \Delta t + \frac{q_3}{3} \Delta t^3$$

$$Q_{55} = q_5 \cdot \Delta t$$

and the coefficients $a = s_k \cdot \cos\alpha_k$, $b = \sin\alpha_k$, $c = -s_k \cdot \sin\alpha_k$, and $d = \cos\alpha_k$. The spectral densities q_1, q_2, q_3, q_4 and q_5 correspond to spectral densities of the state parameters of north coordinate component, east coordinate component, heading rate, heading, and speed, respectively.

The Kalman filter recursively estimates both the state vector parameters $\mathbf{x}_k$ and the error covariance matrix of the state vector parameters $\mathbf{P}_k$. The latter corresponds to the estimated accuracy of the state vector estimate. The calculations are done in two steps, known as the *predict phase* and *update phase*. The predict phase estimates $\mathbf{x}_{k+1}$ and $\mathbf{P}_{k+1}$ based on the system process model and the previous state of the system (i.e., $\mathbf{x}_k$ and $\mathbf{P}_k$). The update phase adjusts these predicted values based on the observation model and the most recent observations (i.e., $\mathbf{z}_{k+1}$). The outputs of the predict and update phases are denoted with the superscripts "−" and "+," respectively.

Thus, the predict phase consists of the following calculations:

$$\mathbf{x}_{k+1}^{-} = \mathbf{\Phi}_k \mathbf{x}_k^{+}$$
$$\mathbf{P}_{k+1}^{-} = \mathbf{\Phi}_k \mathbf{P}_k^{+} \mathbf{\Phi}_k^{T} + \mathbf{Q}_k \tag{5.17}$$

Note that initial values for the state vector and covariance matrix are needed to initialize the Kalman filter. These can be estimated from an absolute positioning system, if available. For example, if the Kalman filter is initiated for a user moving from outdoors (where GNSS is available) to indoors, then the initial position, speed, and heading can be provided by the GNSS receiver.

Next, the outputs of the predict phase are used as inputs to the update phase, as follows:

$$\mathbf{K}_{k+1} = \mathbf{P}_{k+1}^{-} \mathbf{H}_{k+1}^{T} \left[\mathbf{H}_{k+1} \mathbf{P}_{k+1}^{-} \mathbf{H}_{k+1}^{T} + \mathbf{R}_{k+1} \right]^{-1}$$
$$\mathbf{x}_{k+1}^{+} = \mathbf{x}_{k+1}^{-} + \mathbf{K}_{k+1} \left[\mathbf{z}_{k+1} - \mathbf{H}_{k+1} \mathbf{x}_{k+1}^{-} \right] \tag{5.18}$$
$$\mathbf{P}_{k+1}^{+} = \mathbf{P}_{k+1}^{-} - \mathbf{K}_{k+1} \mathbf{H}_{k+1} \mathbf{P}_{k+1}^{-}$$

The matrix $\mathbf{K}_{k+1}$ is known as the *gain matrix*, and it represents the level of "importance" that should be given to the measurements (i.e., observations) when updating the state vector. It is a function of the uncertainty of the state vector after the predict phase (i.e., $\mathbf{P}_{k+1}^{-}$) and the covariance of the observation noise. When the uncertainty of the state vector is high, and the observation noise is low, more emphasis is placed on the most recent measurements. In the limit where the uncertainty of the state vector is zero or the error in the measurements is infinite, the update phase of the Kalman filter has no effect (i.e., $\mathbf{x}_{k+1}^{+} = \mathbf{x}_{k+1}^{-}$, $\mathbf{P}_{k+1}^{+} = \mathbf{P}_{k+1}^{-}$).

The Kalman filter can also be understood as a weighted average of the state vector resulting from the system process model and the state vector resulting from the observation model (using the most recent observation). The result is a better estimate than if either the system process model or the observation model were used alone.

5.4 VISUAL-BASED AND VISUAL-AIDED POSITIONING

Visual positioning solutions can be divided into two categories: visual-based solutions and visual-aided solutions. A visual-based solution outputs positions directly based on information obtained from images, while a visual-aided solution cannot output positions directly, but it can deliver observables that can be integrated with other sensor measurements to obtain a positioning solution. Visual positioning has attracted growing attention in the research community and industry due to the widespread use of the mobile device cameras. Cameras are now becoming standard hardware components in mobile devices. The quality of the images taken from mobile device cameras is improving as well. For example, the Nokia Lumia 1020 smartphone is equipped with a 41-megapixel camera. As a positioning sensor, the camera has the following advantages:

- The image is not affected by magnetic perturbations in the positioning environment;
- Drift is not an issue for cameras.

The main disadvantages are the following:

- The image quality is directly affected by the lighting conditions of the positioning environment;
- The mobile device must be held by the user in a relatively fixed orientation with respect to the body.

Therefore, in some environments or in some user scenarios use of the camera for positioning may not be possible.

5.4.1 Visual-Based Positioning

For a visual-based positioning solution, the traditional approach is to match a query image with georeferenced images stored in a database. Whenever a match is found, the position of the mobile device is considered to be the location associated with the reference image. Therefore, it is an absolute positioning technology. The most common image-matching approach is to match SIFT features [12], which are invariant under image translation, scaling, and rotation. They are also relatively constant under illumination changes [13]. Algorithmically, the image-matching solution is similar to the WLAN fingerprinting solution.

In addition to image-matching, the camera pose can also be estimated using georeferenced landmarks [14]. The three-dimensional locations of the georeferenced landmarks are stored in a database. By matching the SIFT features of a query image with those in a reference image, the three-dimensional landmarks can be recognized. Using these known three-dimensional locations and their two-dimensional coordinates in the query image, the camera pose can be estimated using a technique called *photogrammetric space resection*. This procedure requires known coordinates of at least three landmarks defining a unique plane (i.e., they must not all be in a straight line).

5.4.2 Visual-Aided Positioning

In addition to the above techniques, consecutive images taken by the mobile camera can be used to derive the heading change as well [15]. For heading change detection, the camera serves as a "visual gyroscope," although the measurement rate is much lower than that of a traditional gyroscope sensor due to the heavy image processing that is required. The visual gyroscope can be considered a soft sensor, and it does not require additional hardware other than the camera. It also does not require *a priori* information about the positioning environment, unlike the methods described in Section 5.4.1.

The principle upon which the visual gyroscope is based on the following: When a three-dimensional scene is projected to a two-dimensional image, straight lines are preserved but not the angles between them. In particular, this means parallel lines in the real world will intersect in the image plane. For example, the two parallel lines of a long corridor will intersect to a point in the image plane as shown in Figure 5.6 (the solid dark circle).

Figure 5.6 The vanishing point. The horizontal parallel lines intersect at a vanishing point shown in this figure as a dark solid circle.

Parallel lines in three orthogonal directions will form three intersection points. These intersection points are called *vanishing points*. The vanishing point in the direction of motion is called the *central vanishing point*. The relation between the image coordinates of the central vanishing point and the heading change ($\Delta\alpha$) and the pitch angle (ϕ) can be expressed as [15]:

$$\begin{bmatrix} x \\ y \\ 1 \end{bmatrix} = \begin{bmatrix} f_x \sin(\Delta\alpha) + u\cos(\phi)\cos(\Delta\alpha) \\ -f_y \sin(\phi)\cos(\Delta\alpha) + v\cos(\phi)\cos(\Delta\alpha) \\ \cos(\phi)\cos(\Delta\alpha) \end{bmatrix} \tag{5.19}$$

where (x,y) are the image coordinates of the central vanishing point, f_x and f_y are based on the focal length of the camera expressed in pixels in the x and y directions of the image frame, and (u,v) are the coordinates of the principal point in the image frame. The camera focus lengths (f_x, f_y) and coordinates of the principal point (u,v) are known parameters obtained by camera calibration. The image coordinates (x,y) of the central vanishing point can be determined by image processing techniques, which find the intersection point of parallel straight lines in the direction of motion (e.g., the parallel lines of a long corridor) [15].

Having determined the image coordinates of the central vanishing point, we can derive the heading change and the pitch angle with:

$$\Delta\alpha = \arcsin\frac{x-u}{f_x}$$
$$\phi = \arcsin\frac{y-v}{-f_y\cos(\Delta\alpha)} \tag{5.20}$$

The heading rate can be easily estimated with

$$\dot{\alpha} = \frac{\Delta\alpha}{t_2 - t_1} \tag{5.21}$$

where t_1 and t_2 are the time stamps of the first and second images, respectively. More details of this solution can be found in [15].

5.4.3 Hybrid Integration

The positions derived from the visual-based positioning solutions and heading changes derived from the visual-aided positioning solution can be easily integrated with other sensor measurements and RF signals using the Kalman filter approach described in Section 5.3. This forms a hybrid solution integrating GNSS, motion and orientation sensors, RF signals, and camera images, and it is applicable to a variety of positioning scenarios where use of only a single positioning technology would not be adequate.

5.5 SUMMARY

This chapter discussed three different hybrid positioning solutions appropriate for mobile devices: the PDR solution; the multisensor, multisignal integration solution; and a solution integrating GNSS receiver, sensors, RF signals, and a camera. The PDR solution and the loosely coupled integration scheme based on a Kalman filter were presented in detail. These solutions are simple approaches with great flexibility for integrating heterogeneous measurements from different sources.

References

[1]   Fang, L., et al., "Design of a wireless assisted pedestrian dead reckoning system–the NavMote experience," *IEEE Transactions on Instrumentation and Measurement,* 54, 2342–2358, 2005.

[2]   Beauregard, S., and H. Haas, "Pedestrian dead reckoning: a basis for personal positioning," *Proceedings Of the 3rd Workshop on Positioning, Navigation and Communication, Hannover, Germany,* 2006.

[3]   Weimann, F., and G. Abwerzger, "A pedestrian navigation system for urban and indoor environments," *Proceedings of ION GNSS 20th International Technical Meeting,* Fort Worth, TX, USA, 2007.

[4]   Levi, R., and T. Judd, "Dead reckoning navigational system using accelerometer to measure foot impacts," United States Patent, No. 5,583,776, 1999.

[5]   Grejner-Brzezinska, D., C. Toth, and S. Moafipoor, "Pedestrian tracking and navigation using an adaptive knowledge system based on neural networks," *Journal of Applied Geodesy,* 1, 111–123, 2007.

[6]   Chen, R., P. Ling, and Y. Chen, "A smart phone based PDR solution for indoor navigation," *In Proceedings of ION GNSS 2011 Conference,* Portland, Oregon, USA, September 20-23, 2011. 1404-1408, 2011.

[7]   Renaudin, V., S. Melania and G. Lachapelle, "Step length estimation using handheld inertial sensors," *Sensors,* 12, no. 7: *8507-8525,* 2012.

[8]   Cho, S., and C. Park, "MEMS based pedestrian navigation system," *Journal of Navigation,* 59, 135–153, 2006.

[9]   Ma, J., et al., "Indoor localization based on magnetic anomalies and pedestrian dead reckoning," *In Proc. ION GNSS 2013,* pp. 1033 -1038, 2013.

[10] Kuusniemi, H., Y. Chen, and L. Chen, "Multi-Sensor Multi-Network Positioning," In Chen, R. (Ed.), *Ubiquitous Positioning and Mobile Location-Based Services in Smart Phones,* DOI: 10.4018/978-1-4666-1827-5.ch005, Pennsylvania, U.S., IGI-Global, 97-129, 2012.

[11] Stephen, J., *Development of a Multi-Sensor GNSS Based Vehicle Navigation System* (Master's thesis, University of Calgary, Calgary, Canada). Retrieved from http://www.ucalgary.ca/engo_webdocs/GL/00.20140.JStephen.pdf, 2000.

[12] Lowe, D., "Distinctive image features from scale invariant key points," *Int. J. Comput. Vision* 2004, 60, 91–110, 2004.

[13] Zhang, W., and J. Kosecka, "Image based localization in urban environments," *In Proceedings of the Third International Symposium on three-dimensional Data Processing, Visualization and Transmission*, University of North Carolina, Chapel Hill, NC, USA, 14–16 June 2006.

[14] Li, X., J. Wang, and T. Li, "Seamless positioning and navigation by using geo-referenced images and multi-sensor data," *Sensors 2013.* 13, 9047-9069; doi:10.3390/s130709047, 2013.

[15] Ruotsalainen, L., et al., "A Two-dimensional pedestrian navigation solution aided with a visual gyroscope and a visual odometer," *GPS Solutions.doi 10.1007/s10291-012-0302-8,* 17:575–586, 2013.

Chapter 6

Mobile GIS

According to Esri, one of the leading providers of GIS software, a GIS "integrates hardware, software, and data for capturing, managing, analyzing, and displaying all forms of geographically referenced information." GIS allows us to manipulate geospatial data in a myriad of ways that help us to understand the underlying structure and form conclusions based on the data. GIS is now becoming a powerful tool for urban planning, decision-making, environmental protection, and LBSs.

Modeling, processing, and visualizing geospatial data is a computationally demanding task. Such tasks have traditionally been performed in high-performance servers or desktop PCs. However, the rapid development of mobile technologies is now changing the role of mobile devices in geospatial computing. The reasons that mobile devices are becoming a significant platform for GISs include 1) their mobility, enabled by their small size and wireless connectivity, 2) their positioning capabilities, 3) their increasing computing capabilities, 4) the abundant availability of mobile devices, and 5) good support for developing application software in mobile devices. Mapping applications such as Google Maps and Google Earth available on mobile platforms are now well-known to the general public. The mobile platform is not only making a significant contribution to geospatial information science, but it is also extending the application scope of conventional GIS (e.g., crowdsourcing of geospatial data).

Li and Brimicombe addressed this topic from the perspective that mobile GIS is "an evolution of conventional GIS to being available on wireless mobile devices" [1]. We address this topic from the perspective that mobile GIS is a result of the mobility, positioning capability, and mobile computing capability of mobile devices.

This chapter covers the following four topics: 1) mobile devices as a platform for GIS (Section 6.1), 2) crowdsourcing geospatial data using mobile devices (Section 6.2), 3) ArcGIS for mobile (Section 6.3), and 4) emerging mobile GIS applications (Section 6.4). Section 6.1 introduces the capabilities of mobile devices in terms of mobility, localization, mobile computing, and support for application software development. Section 6.2 addresses how geospatial data can be sourced through public participation in data collection using mobile devices. Section 6.3 concentrates on ArcGIS for mobile, which is a platform targeted for people who have previous GIS education and training. It will explain how to 1) use the ArcGIS mobile application for data acquisition using a mobile device and 2) how to create a map-based application with the ArcGIS Runtime SDKs. The chapter concludes with Section 6.4, which focuses on emerging mobile GIS applications, enabled by geotagged photos and GNSS traces.

6.1 MOBILE DEVICES AS A PLATFORM FOR MOBILE GIS

As discussed in Chapter 1, there are many categories of mobile devices. In this chapter, we will consider high-end mobile devices, known as *smart mobile devices*. These devices can be contrasted from low-end devices, such as feature phones. They offer 1) more computational power, 2) larger memory capacity, 3) larger data storage space, 4) more hardware components and sensors, and 5) better wireless connectivity. Furthermore, a platform of smart mobile device typically have a software development kit (SDK) to support the development of third-party applications, whereas feature phones typically run embedded applications in the form of proprietary firmware developed by the phone manufacturer. Table 6.1 shows the major hardware components and communication capabilities of smart mobile devices. Three mobile devices from three different manufacturers were selected and listed in Table 6.1 for the purposes of comparison. Each mobile device runs a different operating system.

Table 6.1
Typical Hardware Components in Smartphones (www.gsmarena.com)

		iPhone5	Galaxy S4	Lumia 1520
Operating system		iOS 7.0.4	Android 4.2.2	Windows Phone 8
Communication with 4G/LTE		Yes	Yes	Yes
Screen size		4.0 inches	5.0 inches	6.0 inches
Screen resolution		640x1136	1080x1920	1080x1920
Memory		1 GB	2 GB	2 GB
Storage space		64 GB	32 GB	32 GB
CPU speed		Dual-core 1.3 GHz	Quad-core 2.3 GHz	Quad-Core 2.2 GHz
	GNSS	x	x	x
	Accelerometer	x	x	x
Built-in Sensors	Gyro	x	x	x
	Digital compass	x	x	x
	Barometer	-	x	-
	Camera Resolution	8 MP	13 MP	20 MP

The hardware components of a smart mobile device can be divided into three groups: the computational hardware components, the built-in sensors, and the RF technologies, as shown in Figure 6.1. The high-performance computational hardware components increase the capabilities of mobile computing, the built-in sensors enable the capability of positioning, and the RF technologies allow them to be used ubiquitously (i.e., making mobile devices highly portable). As mobile technologies have been developing rapidly in recent years, these components will be continuously enhanced as time goes by.

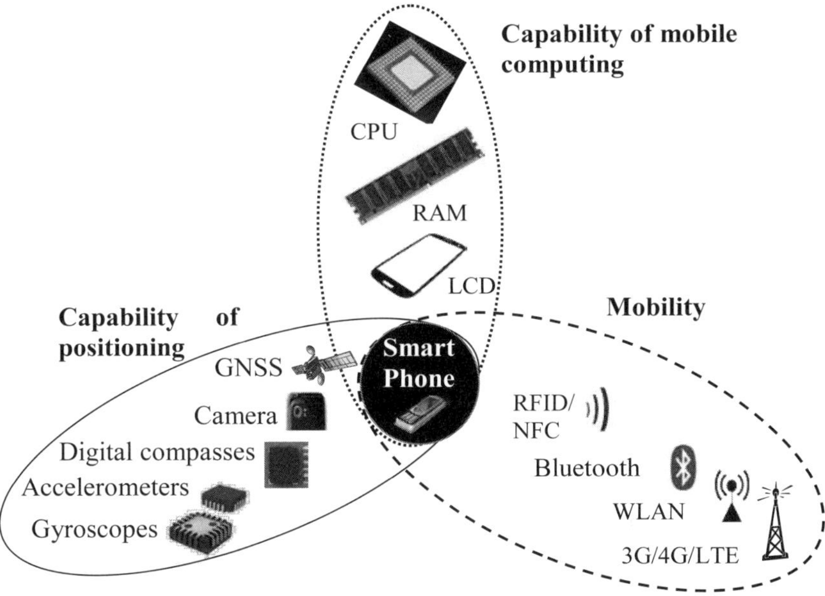

Figure 6.1 Hardware components contributing to the mobility, positioning capability, and mobile computing capability of mobile devices.

6.1.1 Mobile Communications

Typical radio technologies implemented in mobile devices include cellular radio technologies such as 2G, 3G, and 4G long-term evolution (LTE), as well as short-range RF technologies such as WLAN and Bluetooth, as shown in Figure 6.1. In addition to these common radio technologies, some mobile devices support radio frequency identification (RFID) and near field communication (NFC) radio interfaces as well.

The cellular networks offer services with global coverage, while WLAN networks offer local services for small areas. The cellular networks provide services for both voice communication and high-speed wireless Internet access, while the WLAN networks mainly offer wireless Internet access. Of course, voice communication can be implemented using services such as Skype over a WLAN Internet connection.

Such wireless technologies allow these devices to be used in many different environments. For outdoor environments, mobile devices are typically connected to the Internet via cellular connections. The data transmission capacity can be as high as 50 Mbps for the uplink and 150 Mbps for the downlink. For indoor environments, the wireless connectivity is further enhanced by WLAN or Bluetooth technologies. WLAN Internet access is available in most public environments, such as airports, libraries, shopping malls, and restaurants. It is also common in most home environments. Furthermore, WLAN connections are free of charge in many public environments. This widespread wireless connectivity enables access to the Internet ubiquitously.

6.1.2 Mobile Computing Capabilities

The computing capabilities of mobile devices are enabled by the computational hardware components, including central processing units (CPUs), random access memory (RAM), graphic acceleration components, and nonvolatile memory. As geospatial computing is a computationally demanding task, these components play an important role in delivering geospatial data processing results in a timely manner to users.

Other hardware components play an essential role in GIS applications as well, such as high-resolution liquid crystal display (LCD) with touchscreen support and high-quality digital camera, including high-definition (HD) video capability. The touchscreen is essential for visualizing and interacting with geospatial data, and the camera can be used to capture multimedia data for use in GIS applications. In particular, a high-resolution LCD with touchscreen support is essential for 1) displaying maps with good quality, and 2) manipulating the maps comfortably on a small screen using figure gestures for panning, zooming, and rotating.

6.1.3 Positioning Capabilities

The positioning capabilities of mobile devices have been covered extensively in Chapters 1–5, but we will briefly summarize these capabilities below. Sensors integrated into mobile devices that are used for positioning typically include a GNSS receiver, other RF receivers, accelerometers, gyroscopes, and magnetometers, as described in Chapters 2–5. In addition to these common sensors, some mobile devices have additional sensors, such as barometers and ambient light sensors, as shown in Table 6.1. These sensors are enablers for locating mobile devices indoors and outdoors. For outdoor environments, a GNSS receiver is capable of providing a positioning accuracy of 5–10m [2, 3]. For challenging GNSS environments (i.e., where an open view to sky is not available) the positioning accuracy will be degraded significantly or a position solution may not be available at all. Additional measurements from other sensors and signals of opportunity are needed in these cases. In addition to motion and orientation sensors, cameras in mobile devices can also be adopted for positioning [4, 5]. Positioning with visual sensors is not affected by the magnetic perturbations induced by ferrous materials and electronic devices located in the positioning environment. However, visual-based positioning solutions require good lighting conditions in order to capture high-quality images. In addition to the above mentioned methods, WLAN and Bluetooth RSSI measurements have been used for positioning indoors as well. These methods rely on the fact that radio signal strength can be associated with a distance between a transmitter and a receiver [6, 7]. Details on mobile positioning techniques can be found in Chapters 2–5.

6.1.4 Support for Application Development

The two most important software components running inside mobile devices are the operating system and mobile applications. The most common operating systems for mobile devices include the following:

- *iOS* from Apple;
- *Android* from Google;
- *Windows Phone* from Microsoft.

All these operating systems have full support for developing third-party application software, including map-based applications.

Mobile applications (often abbreviated as "apps") are divided into two major categories: native applications and web applications. A native application is built using a SDK for the targeted operating system and runs directly inside the device. Web browsers, calendar, and e-mail apps are typical native applications running in mobile devices. Unlike native applications, web applications are built using web technologies, such as extensible markup language (XML), cascading style sheets (CSS), hyper text markup language (HTML), and JavaScript, as well as many different server-side technologies. Web applications run inside a web browser. Therefore, a connection to the Internet is always needed for running a web application.

Apple's iOS and Google's Android are now the most popular operating systems in terms of attracting a large number of third-party mobile app developers. There are millions of mobile apps available online, and they are predominantly downloaded from application "markets," such as App Store (iOS), Windows Phone marketplace, Google Play Store (Android), and Amazon Appstore for Android.

The programming languages used to create native applications for these operating systems vary. Objective-C is the programming language for iOS, whereas Java is the predominant language for Android (although C and C++ can be used through the Android native development kit). Windows Phone supports a wide range of programming languages, but the preferred languages are C# and Visual Basic (i.e., .NET Framework). Figure 5.1 in Section 5.1 shows a simple map application for an Android mobile device. It is a native application developed with the Android SDK. With this application, users can easily zoom, pan, and rotate the map displayed on the touchscreen using finger gestures. By activating the GPS receiver, the application will center the map at the user's position and automatically pan the map as the user moves. It is not a difficult task to develop such an application because methods for zooming, panning, and rotating maps are already implemented inside the `GoogleMap` class of the Android SDK.

To develop a map application in an Android mobile device, developers need to obtain a Google map API key in order to access the Google map. Details on how to apply for a map API key from Google can be found from the following web page: https://developers.google.com/maps/signup. The API key is a string of characters such as *"BIzaSyATqCdh7_6EMYm-EX-MzQCz3scp2YZxhYc."*

Having obtained the map API key, the application developer needs to add the API key to the `AndroidManifest.xml` file associated with the app as follows:

```
...
<meta-data
    android:name="com.google.android.maps.v2.API_KEY"
        android:value="<your API key>" />
    ...
```

In addition to adding the API key, the `MAP_RECEIVE` and `INTERNET` permissions have to be activated as well. This can be done by adding the following two lines to the `AndroidManifest.xml` file:

```
...
<uses-permission
        android:name="com.example.permission.MAPS_RECEIVE"/>
<uses-permission android:name="android.permission.INTERNET" />
    ...
```

`GoogleMap` is the main class of the map API and is the entry point for all methods such as zooming, panning, and rotating maps. The `GoogleMap` object cannot be instantiated directly; rather, it must be obtained from the `getMap()` method of a `MapFragment` or `MapView` object as shown in the following example code:

```
...
private GoogleMap map;
map = ((MapFragment) getFragmentManager().findFragmentById(R.id.map)).getMap();
map.moveCamera(CameraUpdateFactory.newLatLng(new LatLng(27.715523,
```

-97.328658)); //center the map to this location.

...

For more details on developing a map application, please refer to the web page https://developers.google.com/maps/documentation/android/.

6.2 CROWDSOURCING GEOSPATIAL DATA USING MOBILE DEVICES

Many different geospatial sensors have been adopted for mapping the Earth's surface, including sensors operating in space [8], in the air [9], and on the ground [10]. The number of commercial and governmental geospatial data acquisition systems continues to increase, and the amount of data these systems collect is enormous. Despite this enormity of data, the current geospatial sensor technologies cannot achieve the spatial and temporal resolutions needed for mapping all changes occurring on the Earth's surface, especially concerning small-scale details and fast-evolving situations (e.g., during a natural hazard event such as a forest fire or a hurricane). The concept of "citizen as sensors" has attracted the attention of scientists in geospatial information science because of the abundance of mobile devices. The size of the current citizen sensor network is more than 6 billion, and this number is increasing all the time [11]. By connecting them all to a cloud computing system, the potential of the citizen sensor network to crowdsource geospatial data is only limited by our imagination. As explained in Chapter 5, mobile devices are now equipped with a GPS receiver and other motion and orientation sensors and thus offer an ideal platform for crowdsourcing geospatial data due to their abundant availability, mobility, capability of positioning, and capability of mobile computing. Crowdsourced data can provide valuable information for observing and analyzing the rapid changes of the Earth's surface. This also offers the general public a unique opportunity to make contributions in mapping their own neighborhoods.

Figure 6.2 shows part of a GNSS trace recorded by a mobile device. The trace is a driving path of about 15 minutes. Information such as time stamps, vehicle positions, and vehicle elevations are logged every second to a file in GPS exchange (GPX) format. The logged information can be used to derive parameters such as driving speed and driving heading. This information can be further used for extracting realtime traffic information, identifying road construction sites, and discovering new road segments if we have sufficient crowdsourcing users to contribute their GNSS traces in realtime to the data processing center.

For example, we can identify a road construction site (or car accident site) when there is no vehicle passing through a particular road segment for a certain period when we would normally expect there to be. We can only make such a conclusion, however, based on validation from a large amount of crowdsourcing users.

Figure 6.2 A part of a GNSS trace of a driving path logged with a mobile device. Information such as time stamps, vehicle positions, and elevations are logged every second to a file in GPX format. The figure on the left is overlaid with a driving path, while the one on the right shows the original map from OpenStreetMap. It is obvious that a new road segment has been added to the end of the driving path as shown in the picture on the left.

The situation is similar for deriving traffic information. Although the speeds and headings of a single GNSS trace can provide a useful patchwork of traffic information, we still need a large amount of such GNSS traces in order to create a realtime traffic map covering a large city.

GNSS traces can be used to identify new road segments as well. The screen shot on the left of Figure 6.2 shows that a new road segment was identified at the end of the driving path. This finding can be further confirmed if there are multiple GNSS traces available for the same area.

6.2.1 The GPX Data Format

The most common data format for logging a GNSS trace is the GPX format (http://en.wikipedia.org/wiki/GPS_eXchange_Format), which is a file format following an XML schema. It can be used to describe waypoints, tracks, and routines. It timestamps the positions and elevations for each waypoint as shown in the following:

```
...
<trkpt lat="27.714736" lon="-97.328259">
<ele>-22</ele>
<time>2013-11-12T01:16:03.322Z</time>
</trkpt>
...
```

The GPS coordinates are given in the WGS-84 coordinate system. The latitude and longitude are in degrees decimal, while the elevation is in meters. The time stamp is in coordinated universal time (UTC). The driving heading and driving speed can be derived from coordinates and time stamps of two consecutive waypoints. GPX is an open standard, therefore, it can be used without any license fees. Most location- and map-based software such as Google Earth, Google Maps, and ArcGIS support the GPX format. It is easy to import GNSS traces to these applications.

6.2.2 *OpenStreetMap*

OpenStreetMap (http://www.openstreetmap.org) is a very good example of demonstrating the potential of crowdsourcing geospatial data using mobile devices. It is an initiative to create and provide free geographic data, such as street maps, to the public. It is built by a community of mappers who contribute and maintain the data from all over the world. *OpenStreetMap* is dedicated to encouraging the growth, development, and distribution of free geospatial data for all. Currently, the total number of registered mappers is 1,503,241, while the total number of uploaded GPS points is 3,805,065,547. These numbers are growing all the time and are available at http://www.openstreetmap.org/stats/data_stats.html.

The data sources for *OpenStreetMap* include 1) GNSS traces; 2) local knowledge such as names of the streets, roads, shops, and buildings; and 3) aerial and satellite images. GNSS traces are currently the most common data source collected by the community. To contribute data to the *OpenStreetMap*, users need to do the following:

- Record GNSS traces using their mobile devices. The GNSS traces are typically saved to a file in GPX format;
- Upload the GNSS traces to the *OpenStreetMap* database;

- Edit the GNSS traces using the map editor [e.g., the Java OpenStreetMap (JOSM) editor].

More information on how to add data to *OpenStreetMap* can be found on the web page: http://wiki.openstreetmap.org/wiki/Beginners_Guide.

One issue with crowdsourced geospatial data is quality assurance. Traditionally surveying and mapping tasks are carried out by trained professionals. They follow well-defined specifications and procedures for quality assurance during data collection. Contributors to crowdsourced data may not have the needed education and training to follow these professional specifications and procedures. It would be unrealistic to expect that the quality of the data obtained from crowdsourcing is similar to that collected by professionals. This issue has been investigated by scientists, and it was determined that the average geometric accuracy of the data obtained from crowdsourcing is about 6m [12, 13]. A discrepancy of about 5–10m between the GNSS trace and the road maps is also clearly visible from Figure 6.2.

6.3 ArcGIS FOR MOBILE

According to Esri, the provider of ArcGIS, it is "a comprehensive system that allows GIS professionals to collect, organize, manage, analyze, and distribute geographic information." The system consists of a set of software including ArcGIS for Server, ArcGIS for Desktop, ArcGIS Online, and ArcGIS for Mobile. ArcGIS for Server offers organizations possibilities to provide secure and reliable GIS services for various web, mobile, and desktop applications. ArcGIS for Desktop is a desktop application that allows GIS users to perform spatial analysis, data management, map visualization, and map editing. ArcGIS Online is a cloud-based platform that allows organizations to use Esri's cloud service to create, share, and access their maps and other geospatial contents including authoritative basemaps published by Esri. ArcGIS for Mobile consists of ArcGIS applications for mobile devices and Runtime SDKs for developing third-party GIS applications and LBSs. The mobile ArcGIS applications and SDKs are available for all three major mobile operating systems including iOS, Android, and Windows Mobile.

Figure 6.3(a) shows a screenshot of the ArcGIS mobile application running in an Android phone. The application is developed by Esri and can be downloaded freely from software application markets such as App Store, Windows Phone marketplace, Google Play Store, and Amazon Appstore for Android. Mobile users can utilize the ArcGIS mobile application to display and navigate maps; to find addresses and places; to identify locations and GIS features; to measure lines and areas; to find and to share maps from ArcGIS Online; and to collect GIS data (e.g., using the GPS in mobile devices).

Figure 6.3(b) shows a GNSS trace of a walking path recorded by the ArcGIS application. The mobile user just needs to push an icon on the touchscreen to record a GPS position while he/she is walking. The cumulated length of the walking path is also calculated and displayed on the screen. The user can upload the recorded walking path to an ArcGIS online server, a corporate ArcGIS server, or a personal ArcGIS portal from the field immediately after the surveying. The new geospatial objects will be visible in realtime to authorized online users (for example, a project manager working in an office). This function allows multiple people to collect data in the field simultaneously and aggregate all data to a server in realtime. This is a very good example demonstrating the potential created by the mobility, the locating capability, and the mobile computing capability of mobile devices for mobile GIS applications.

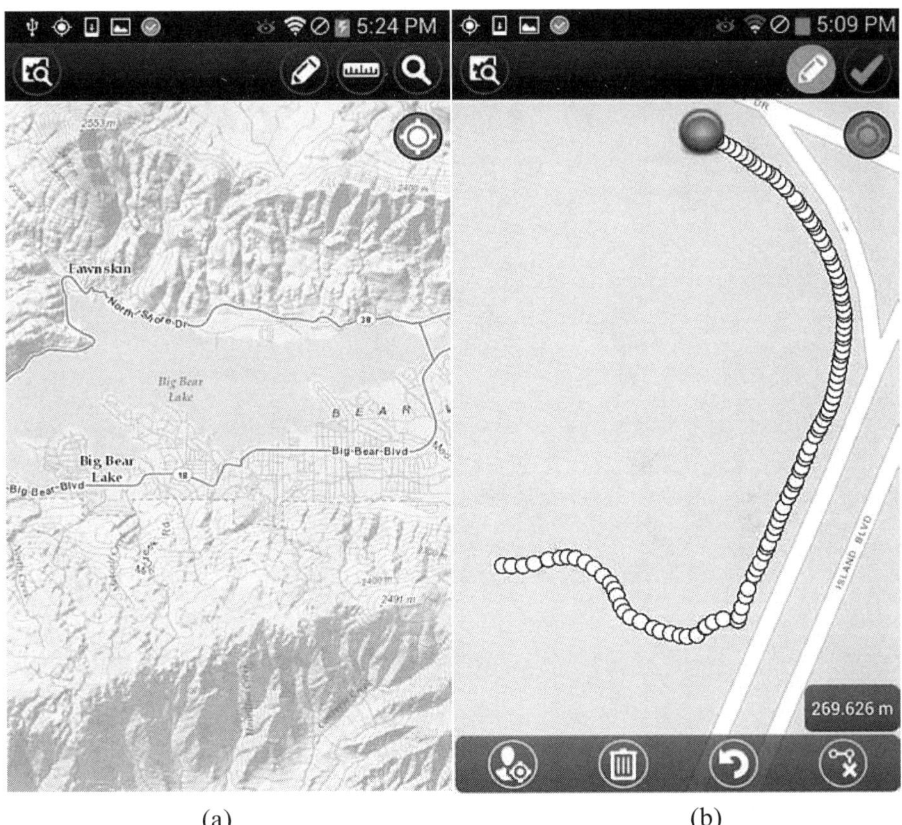

(a)   (b)

Figure 6.3 Screenshots of a walking path recorded using ArcGIS for Android. Each circle represents a tracking point. The user can push the bottom left icon to record the current GPS position as a tracking point. The total length of the walking path is also shown on the screen (269.626m).

In many ways, the ArcGIS mobile application meets the needs for professional GIS work. In addition to data acquisition, the mobile user can add, update, and delete features of geospatial objects using the application. Similar applications are also available for the other two mobile operating systems: iOS and Windows Phone.

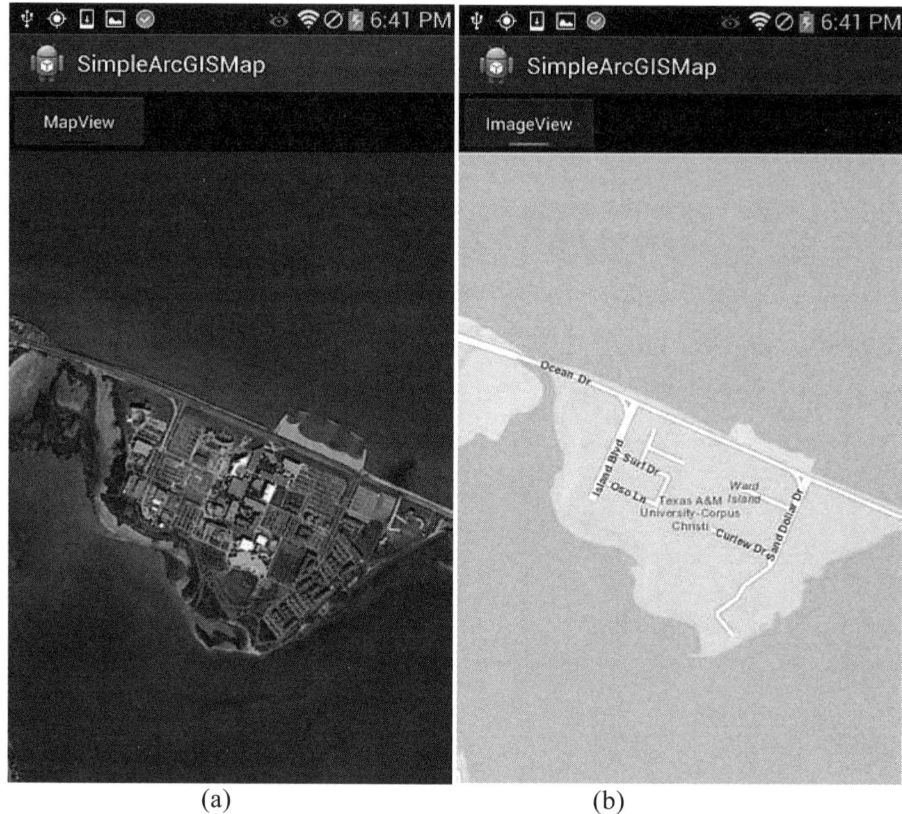

(a)   (b)

Figure 6.4 A simple map application developed with the ArcGIS Runtime SDK for Android. The image view is shown on the left, while the map view is on the right.

In addition to the ArcGIS applications, the ArcGIS Runtime SDKs are essential resources for developing mobile GIS applications. These SDKs offer many GIS functions, such as the following:

- Support for different coordinate systems;
- Map projections;
- Spatial analysis;
- Data manipulation;
- Data management.

These functions support the development of professional GIS applications. Figure 6.4 shows two screenshots of a simple map application developed for an Android phone using the ArcGIS Runtime SDK. The user can easily switch to an image view (left) or to a map view (right). The zooming and panning operations for the maps are implemented inside the `MapView` object of the SDK. Table 6.2 shows the source code needed for implementing the example shown in Figure 6.4.

Developing a map application for mobile devices using ArcGIS Runtime SDK is quite similar to that of developing a Google Map application. One difference, however, is that no API key is needed for ArcGIS Runtime SDKs. The user needs only to install the ArcGIS Runtime SDK as a group of plug-ins to an Android integrated development environment (IDE) such as Eclipse. More information about installing and using the ArcGIS Runtime SDK can be found from the web site: https://developers.arcgis.com/en/android/install.html.

6.4 EMERGING MOBILE GIS APPLICATIONS

If one searches from Google Play Store, the software market for Android applications, with keywords "location," "map," "navigation," or "GIS" he or she will likely find hundreds, if not thousands, of relevant applications. The large number of mobile users and the availability of SDKs play essential roles in supporting the development of mobile map and GIS applications.

Enabled by its mobility, positioning capability, and mobile computing capability, the mobile device is a powerful platform to host innovative GIS applications including conventional GIS applications, such as ArcGIS for mobile, and other new applications that extend the scope of the conventional GIS applications.

There is considerable diversity in mobile GIS applications. Also, there is a diminishing distinction between mobile GIS applications and LBSs (covered in Chapter 7), so it can be difficult to categorize all such applications into mutually exclusive groups. For example, in some cases geotagging photos can be considered a GIS-related activity, whereas in other cases this is a feature of an LBS. Despite these difficulties, we categorize mobile GIS applications into the following three groups:

- ArcGIS applications for mobile devices;
- Geotagged photos using mobile devices;
- Collecting waypoints and tracks using mobile devices.

Typical users of ArcGIS mobile applications are GIS professionals or people with previous GIS education and training. The typical users of the applications in the second and third groups are people who do not have GIS education and training. These applications, nonetheless, are opening a new landscape for GIS applications.

Table 6.2

Source Code for Implementing a Simple Map Application Using the ArcGIS Runtime SDK for Android

```
public class SimpleArcGISMapActivity extends Activity {

        MapView mMapView;
        ArcGISTiledMapServiceLayer mapLayer;
        ArcGISTiledMapServiceLayer imageLayer;

        public void onCreate(Bundle savedInstanceState) {
        super.onCreate(savedInstanceState);
        setContentView(R.layout.main);

        mMapView = new MapView(this); // Create the MapView object

        imageLayer = new
        ArcGISTiledMapServiceLayer("http://services.arcgisonline.com/ArcGIS/rest/services/
        World_Imagery/MapServer"); // Create the map layer
        mapLayer = new
        ArcGISTiledMapServiceLayer("http://services.arcgisonline.com/ArcGIS/rest/services/
        World_Street_Map/MapServer"); // Create the image layer

        mMapView = (MapView)findViewById(R.id.map);
        // Add dynamic layer to MapView
        mMapView.addLayer(imageLayer);
        }

        public void onToggleClicked(View view) {
        boolean on = ((ToggleButton) view).isChecked();

        if (on) {
                mMapView.addLayer(mapLayer); // Switch to map layer
                }
        else {
                mMapView.addLayer(imageLayer); // Switch to image layer
                }
        }
    ...
}
```

6.4.1 Geotagged Photos Using Mobile Devices

Posting geotagged photos to web sites, such as Facebook (http://www.facebook.com), Panoramio (http://www.panoramio.com), or Flickr (http://www.flickr.com), is a voluntary approach to "mapping" the Earth. It is now possible to embed location information to the header of an image file. This information is not visible in the picture but can be extracted by most applications and used to pinpoint the location of a photo on a map. Currently, the number of photos uploaded to the Flickr web site is 12,565,267,544. This number is based on the index of a photo uploaded on February 16, 2014. Of course, not all of these photos include location information. By the time of January 31 2013, about 200 millions Flickr photos did include location information [14]. This number is now increasing because most smartphone cameras support the function of geotaging images. These large amounts of georeferenced images represent a significant resource of geospatial information!

The software in most mobile devices is capable of writing to the image file the position of the mobile device at the time the image was taken. The mobile device is usually located using its built-in GNSS receiver. When uploading the image file to a web service, the web application, such as Flickr, will automatically extract the geolocation information from the image file and pinpoint the photo to a correct location on the map. In case geolocation information is not found from the image file (e.g., for photos taken with a camera that does not have a GNSS receiver), the web application usually allows users to pick a location from the map based on their own knowledge. This approach, however, is not as convenient for the users, nor is it as accurate as the geotagged photos. Figure 6.5 shows a geotagged photo uploaded to Panoramio and displayed in Google Earth.

Figure 6.5 Example of a geotagged photo uploaded to Panoramio and displayed in Google Earth.

6.4.2 Collecting Waypoints and Tracks Using Mobile Devices

Collecting waypoints and tracks is an easy task using a mobile device. GNSS traces have been very useful for crowdsourcing geospatial data as discussed in Section 6.2. They can be used for other potential applications as well, for example, for 1) measuring driving safety and risk and 2) creating a digital diary that records the footprints of our activities such as hiking, running, biking, and holiday trips.

The use of GNSS to improve driving safety could potentially save many lives. From a technical point of view, it is not a difficult task to assess the driving safety and risks of using GNSS traces because information such as the following can be extracted or derived from GNSS traces:

- Driving times (daytime or nighttime);
- Driving areas (downtown, highway, or suburban);
- Driving speeds;
- Driving headings;
- Driving behaviors such as sudden accelerations and frequent unnecessary lane changes.

GNSS traces are very precise observations for measuring driving safety and risks. For a good driver, the GNSS traces offer good evidence for negotiating a lower insurance fee with a car insurance company. Many issues remain before such practices are put into widespread use, but they are mostly non-technical in nature. For example, some privacy groups have raised objections concerning government plans to place GNSS tracking devices into vehicles.

On a more grassroots level, trajectories from our daily activities can be recorded using mobile devices and analyzed for private use or shared voluntarily through the web. Statistics such as daily walking distances and average walking speeds can be derived from GNSS traces and used to better understand one's personal fitness. Traces from holiday trips offer us a way to preserve memories from these journeys at a whole new level of detail (in conjunction with geotagged photos and video clips).

Figure 6.6 shows two screenshots from the My Tracks application with a walking path on the left and the performance statistics on the right. My Tracks is an open source application for Android devices, whose development is led by Google. It can be used to record GNSS traces of walking, running, and driving. A GNSS trace is saved to a file in the mobile device in GPX format. The traces can be replayed in a mobile device, exported to a file (e.g., in GPX format), or uploaded directly to Google Maps.

Figure 6.6 GNSS trace of a walking path (left) and the corresponding performance statistics (right).

6.5 SUMMARY

The application scope of conventional GIS has extended from desktops to the mobile environment, by virtue of the mobility, the positioning capability, and the mobile computing capability of mobile devices. Mobile GIS is now playing an essential role in geospatial information science. This chapter has addressed the following topics concerning mobile GIS: the mobile device as a platform for mobile GIS; crowdsourcing geospatial data using mobile devices; ArcGIS for mobile; and emerging mobile GIS applications. While GIS technologies have been developing rapidly in many different areas, this chapter focuses on the developments related to mobile devices. It covers applications for GIS professionals and applications for the general public.

References

[1]   Li, Y., L. Pei, and A. Brimicombe, "Mobile Geographic Information Systems," In Chen, R. (Ed.), *Ubiquitous Positioning and Mobile Location-Based Services in Smart Phones*, DOI: 10.4018/978-1-4666-1827-5.ch003, Pennsylvania, IGI-Global, 230-253, 2012.

[2]   Chen, R., "Introduction to smart phone positioning," In Chen, R. (Ed.), *Ubiquitous Positioning and Mobile Location-Based Services in Smart Phones*, DOI: 10.4018/978-1-4666-1827-5, Pennsylvania, IGI-Global, 2012.

[3]   Zandbergen P. A., "Accuracy of iPhone locations: A comparison of assisted GPS, WLAN and cellular positioning," *Transactions in GIS, 2009,* 13(s1): 5-26, 2009.

[4]   Ruotsalainen, R., et al., "A two-dimensional pedestrian navigation solution aided with a visual gyroscope and a visual odometer," *GPS Solutions. doi 10.1007/s10291-012-0302-8*, Volume 17, Issue 4, pp 575-586, 2013.

[5]   Ruotsalainen L., H. Kuusniemi, and R. Chen, "Visual-aided two-dimensional pedestrian indoor navigation with a smartphone," *Journal of Global Positioning Systems (2011).* Vol.10, No.1:11-18. DOI: 10.5081/jgps.10.1.11.

[6]   Chen, R., L. Pei, and H. Lelppäkoski, "WLAN and Bluetooth positioning in smart phones," In Chen, R. (Ed.), *Ubiquitous Positioning and Mobile Location-Based Services in Smart Phones*, DOI: 10.4018/978-1-4666-1827-5.ch003, Pennsylvania, IGI-Global, 44-68, 2012.

[7]   Chen L., et al., "Bayesian fusion for indoor positioning using Bluetooth fingerprints," *Wireless Personal Communications,* Vol. 70, Issue 4, pp 1735-1745. (DOI: 10.1007/s11277-012-0777-1, 2012.

[8]   NASA's Earth Observing System Project Science Office: http://eospso.gsfc.nasa.gov/content/nasas-earth-observing-system-project-science-office.

[9]   Airborne Remote Sensing Measurements: http://www.asdi.com/applications/remote-sensing/airborne-remote-sensing.

[10] Kukko, A., et al., "Road environment mapping system of the Finnish Geodetic Institute – FGI ROAMER," *Proceedings of the ISPRS Workshop on Laser Scanning 2007 and SilviLaser 2007*, Espoo, Finland, 12-14 Sept. 2007, pp. 241-247.

[11] Goodchild, M. F., "Citizens as sensors: The world of volunteered geography," In Dodge M., Kitchin, R., and Perkins, C., *The Map Reader: Theories of Mapping Practice and Cartographic Representation.* John Wiley & Sons, Ltd, Chichester, UK. doi: 10.1002/9780470979587.ch48.

[12] Haklay, M., "How good is volunteered geographical information? A comparative study of OpenStreetMap and Ordnance Survey datasets," *ENVIRON PLANN B*, 37 (4) 682 – 703, 2010.

[13] Heipke, C., "Crowdsourcing geospatial data," *ISPRS Journal of Photogrammetry and Remote Sensing,* Volume 65, Issue 6, pp. 550-557, 2010.

[14] Wider, T., D. Palacio, and R. S. Purves, "Georeferencing Images Using Tags: Applications With Flickr," *16th AGILE Conference on Geographic Information Science,* 14-17 May 2013, Leuven, Belgium.

Chapter 7

Mobile LBSs

In Chapters 5 and 6, we explored the broad class of mobile applications, known as context-aware applications. Now, we will narrow our focus to *location-aware applications* and their associated services, generally known as LBSs. The term LBS has been adopted by the mobile and information technology industries to encompass many types of services where the geographic position of the user plays a central role. As stated in Chapter 7, *location awareness* is a forerunner in the broader development of context awareness, and accordingly, LBS leads the way in terms of market adoption and number of existing applications. In this chapter, we will take a look at how LBS technologies and the resulting LBS market have developed. We then survey the different categories of existing LBSs. Next, we discuss important LBS concepts, such as data types and commonly used functions. Finally, we examine the current market and users' attitudes toward LBS, which are important for any developer to keep in mind when developing a new LBS or seeking to improve an existing one.

7.1 WHAT IS A LBS?

Although the term LBS is widely used, it is yet another term that is ill-defined. In the past, the computer software literature was careful to distinguish between *services* and *applications*, but in the modern context of smartphones and cloud computing, this distinction is increasingly blurred.

One way to maintain this distinction is to think of services as the high-level functionalities created and maintained by a service provider (typically a company or other organization), whereas applications are platform-specific implementations of a service, whether they are implemented as standalone applications or accessed through a general-purpose web browser as a *web application*. In an increasing number of cases, a service provider may expose its service via an API (e.g., a

RESTful web service[15]) and third parties create applications that consume this API. This scenario is depicted in Figure 7.1.

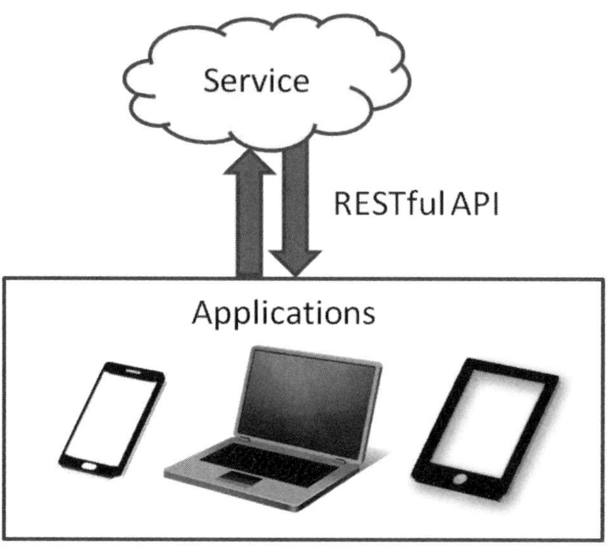

Figure 7.1 Depiction of the difference between a service and an application.

For the remainder of this chapter, we will use the terms service and application in a broad sense (almost interchangeably) since the distinction does not impact our discussion very deeply and because the existing literature is anyways not very consistent in making such distinctions.

With this framework in mind, the term LBS has been used to refer to many different kinds of services/applications that utilize location as an input or those that are somehow related to geography and physical position. Depending on one's perspective, this could include navigation applications, location-centric search applications, or social media applications that identify the current positions of users when they share or create various content (e.g., a location where a photo was taken).

Generally speaking, LBSs of all types are highly "mobile-based," meaning they can be accessed and used from mobile devices (mobile phones, tablets, etc.) and are generally optimized for such access. Typically, they can be accessed as well from traditional PCs, usually through a web interface. In this chapter, we will

---

[15] *RESTful* describes interfaces that follow the architectural style known as representational state transfer (REST). It is today the most prominent style for implementing web services. In HTTP, it is typified by commands such as *GET, POST, PUT,* and *DELETE.*

focus on mobile LBSs, a term that emphasizes the mobility aspect. Given the current proliferation of mobile devices, however, the addition of the word "mobile" to this term is somewhat superfluous. Thus, we will usually drop it and assume it is obvious from the context.

Various authors have offered definitions of LBS (e.g., [1]). One widely cited definition, which has been in use by the Open Geospatial Consortium (OGC) since at least 2003, is the following:

A wireless-IP service that uses geographic information to serve a mobile user. Any application service that exploits the position of a mobile terminal [2].

Despite its longevity, in our view this definition has some weaknesses. Our primary concern is that it is a two-part definition, and it is not clear whether the definition requires an LBS to satisfy both parts or just one of the two. For example, every map application uses geographic information but doesn't necessarily exploit the position of the user. For this reason, we offer a modified version of this definition:

A (mobile) location-based service is any application or service that exploits the position of a mobile device.

One example of a LBS is *Foursquare*, a service that allows users to "check in" to locations. In this way, they can interact with other users who either are currently at the same location or have been there previously. At the same time, the service can be used to find out more information about particular locations, to read reviews and recommendations about locations, and even to get discounts from businesses that are partnered with the service. According to Foursquare, over 40 million people use this service worldwide [3].

Section 9.3 will take a deeper look at the different kinds of LBSs that currently exist. Before doing so, however, it is instructive to take a look at how LBSs came into being.

7.2 HISTORY OF LBSs

The term LBS used in the context of computer technology first came into use in the early 1990s. The first such usage in the literature (per our exhaustive search) was from a January 1993 article titled "The Future Mobility Market," where the author, DeFeo, states that "*...in this new age of personal communication, customers require a full array of information and location-based services*" [4]. He does not explicitly define LBSs, perhaps because the meaning is more or less self-evident within this context.

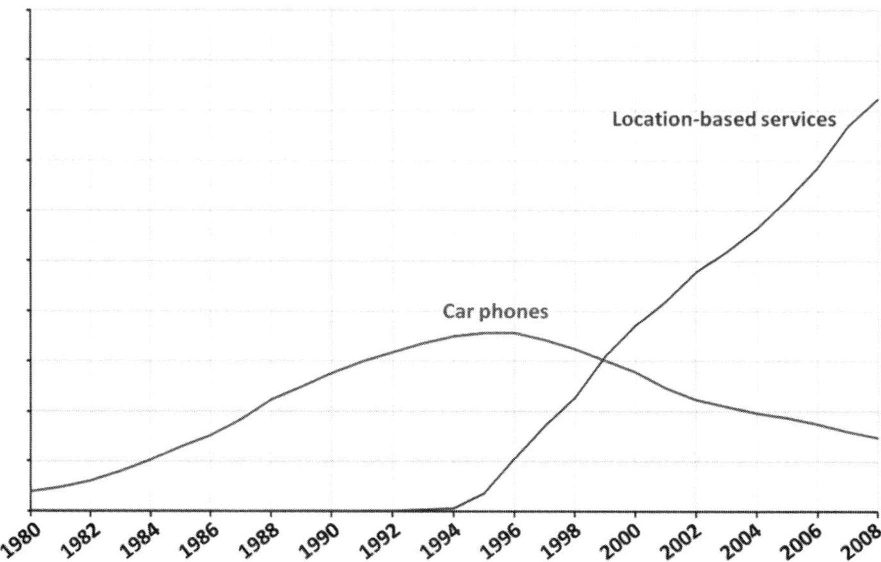

Figure 7.2 Relative prevalence of the terms LBSs and car phones. Data from Google Books Ngrams corpus [5].

In any case, it is shortly thereafter where we see a number of different authors using either this term or a similar one (e.g., location-aware services [6]) to refer to different kinds of services that involve both mobile devices and a location component. A good example can be found in [7], which also offers a clear rationale why LBSs are needed:

Mobility brings about the scenario where the information or service the mobile user needs depends on the current location of the user. However, such a service should be available transparently, without the user having to consciously take into account their changing location. Location based services attempt to provide such transparency where the nature of the service or data provided may differ from one location to another.

Thus, the one common attribute of LBSs is that they should seamlessly and transparently adapt based on the user's geographic position. Other than that, there is a great diversity in the type of LBSs that exist, due to the fact that many functions of daily life are dependent on one's current geographic position.

As a result of this fact, we witnessed a rapid rise in LBSs that closely followed the increase in mobile subscribers. Figure 7.2, produced from the Google Books Ngram Viewer, shows the relative rise of the term LBSs (in the Google Books corpus) during the late twentieth century and early twenty-first century [5]. For comparison purposes, Figure 7.2 also shows the prevalence of the term "car phones," which rose in popularity during the early 1980s but peaked in the mid1990s. Today, of course, most people don't speak of car phones but rather mobile phones (or cell phones) and increasingly *smartphones*. This figure also points toward the longevity of the concept of LBSs. Although the devices we use to access them are changing rapidly, the relevance of LBSs continues to rise. In the following sections, we will take a detailed look at how this rise has taken place.

7.2.1 Three Converging Technologies: Mobile, Internet, Location

What is abundantly clear to us today—and which few had the foresight to recognize at the time—is that the 1990s saw a convergence of several key technologies that made LBSs possible and in some sense necessary. First is the rapid development and consumer uptake of mobile communication devices, starting with cellular phones and PDAs, which were eventually merged into the smartphone. This was not only key to making LBSs possible, but due to the way mobile devices have changed our way of working and living, they have made LBSs more and more necessary in order to "navigate the digital world." Between 1995 and 2000, the number of mobile cellular subscribers per capita (worldwide) grew from 1.6% to 12.1%. By 2010, this figure grew to 77.1%. In some regions, the growth of mobile communications was even more remarkable. For example, in Latin America and the Caribbean, the same figure grew from 0.8% in 1995 to 96.8% in 2010 [8].

Second, LBSs would be nothing without this digital world (i.e., the information supplied by the services themselves). In this sense, the rapid development in the 1990s of Internet-based services in general helped to make LBSs possible. Remember that the web itself was invented between late 1990 and early 1991 [9]. In 1995, the number of Internet users was less than 10% of the population in the United States and less than 1% worldwide [8]. For most people during the early 1990s, LBSs were restricted perhaps to the yellow pages and paper maps—or asking the concierge at the hotel for restaurant recommendations.

Finally and crucially, LBSs would not be possible (or at least not practical) without methods to determine a user's position that do not rely on "manual input." Although a full review of this history is beyond the scope of this chapter, we note that these methods started to become a reality in the early 1990s. This includes the development of GPS [10], which declared full operational capacity (FOC) in April 1995, and to a lesser extent GLONASS in Russia, which achieved its full constellation in December 1995 [11].

In the early 1990s, however, GNSS receivers measured nearly 7×7cm and cost several hundreds of U.S. dollars [12]. Other systems to determine a user's position using satellite signals, such as Geostar and Qualcomm's OmniTRACS, came into existence in the late 1980s and early 1990s, but users were primarily limited to long-haul trucking companies and a few maritime users [13]. As earlier chapters make clear, however, satellite-based positioning is not the only method for obtaining a user's position.

In fact, it was ground-based cellular network operators (and their partners) that led the way in developing consumer-grade positioning capabilities, allowing mass market LBSs to come into being.[16] A major catalyst for this was the adoption of E-911 regulations in the United States, starting in 1996. These regulations require wireless network operators to be able to locate their users in the case of emergency calls. Therefore, emergency calls placed from mobile phones basically fit our definition of a LBS.

The E-911 regulations were implemented in two phases. In the first phase, which was enforced starting in April 1998, network operators had to report "the location of the base station or cell site receiving a 911 call to the [Public Safety Answering Point]" [14]. The second phase, which required greater positioning accuracy, was more complicated because the requirements depended on whether providers opted to use network-based techniques (e.g., ToA and AoA) or handset-based techniques (E-OTD, A-GPS, etc.). Increasing requirements were phased in between October 2001 and December 2005. They required accuracies ranging from 50m (67% of calls) to 300m (95% of calls), depending on the positioning method employed [15]. It should be noted that these requirements apply only to outdoor cases. No positioning accuracy requirements yet exist for 911 calls made from a mobile phone located indoors.

7.2.2 The Birth of Commercial LBS

With the convergence of these three core technologies, we began to see the first examples of commercial LBSs in the mid to late 1990s. Surprisingly, however, this recent history is not well-documented; it can primarily be assembled from various corporate press releases of early LBS companies and archived media sources. The developments can also be seen by examining patent filings from this period concerning related technologies.

One of the earliest patent examples is an automotive-related patent application with the title "Vehicle Locating and Communicating Method and Apparatus," filed in January 1991 by By-Word Technologies, Inc [16]. According to several newspaper articles from the early 1990s, there were a number of small companies offering such services around the same time frame, such as *International Teletrac Systems* [17]. It was only a few years later that General Motors and its partners developed a related service called *OnStar* (originally developed as "Project Beacon" beginning in 1994), which allowed vehicles to be remotely located and connected to services such as driving directions and

---

[16] Some may assume that GPS led the race of mobile positioning technologies, but remember it was not until *after* 1 May 2000, when GPS's selective availability was turned off, that accurate satellite-based positioning would come into commercial (civilian) usage.

emergency response. Introduced to the market in 1996, OnStar can be considered as one of the first widely adopted examples of a commercial LBS [18].

Other early LBS-related patents include "System and methods for remotely accessing a selected group of items of interest from a database," filed in January 1995 by Civix Corporation (granted May 7, 2002), which describes accessing location-related information based on a user's position [19]. Similarly, Apple Computer filed the application "Time and location based computing" in May 1995 (granted June 24, 1997) [20]. By the end of the 1990s, there were dozens of patent applications concerning various aspects of LBSs. Within a few years much of this intellectual property began to make its way into various consumer products and services.

Another vehicle-related commercial LBS that deserves mentioning was the CellPoint tracking and recovery system, which used GSM positioning technology to track stolen vehicles. This service, originally developed by Wasp International, was first operational in South Africa in 1996 [21]. The service later expanded to Europe, and in 1999 it was implemented in Sweden for mobile phones, known as "Tele2Mobil Position" [22]. In this regard, it was an important pioneer in the development of positioning based on standard cellular networks.

With regard to LBS offered for mobile devices, another one of the early players was Palm Computing (now part of Hewlett-Packard), makers of the PalmPilot and other PDAs. PDAs were essentially the precursors of modern smartphones. In North America, starting in May 1999, Palm Computing began selling the Palm VII (a successor to the PalmPilot line of PDAs), which was capable of coarsely determining its position in terms of the cellular base station it was currently using. The Palm OS 3.2 gave application developers access to this position in the form of a zip code, allowing the creation of various LBSs [23]. The Palm VII also came preloaded with applications from *MapQuest* and *The Weather Channel*, providing LBSs such as traffic and weather updates [24].

Another often cited example of an early mobile LBS is *ZagMe*, a location-based advertising service that began operating in the United Kingdom in late 2000 [25]. The service concept was that users would receive promotional text messages to their mobile phones when entering a particular shopping mall. The company responsible for this service did not, however, have mobile positioning technology in place, so users were required to manually activate the service upon entering the mall by calling or sending a text message to a specified number [26]. Thus, it lacked the automatic positioning component of our definition of LBS. Nonetheless, it can be considered a pioneer in mobile location-based advertising, which is today a massive industry.

By the turn of the millennium, many LBSs were either under development or already operating commercially. Apart from the examples discussed above, other major players in the early LBS market included Ericsson, Nokia, Europolitan (now part of Vodafone), NTT Docomo, Telia (now TeliaSonera), and Motorola. Let's fast forward, however, and take a look at the types of LBSs that exist today.

7.3 TYPES OF LBSs

Today LBSs are so ubiquitous and integral to our daily lives that we may not even realize we are using them. In fact, all mobile operators determine the (rough) position of their users' devices in order to deliver the mobile services themselves, so they are by our definition a kind of LBS. We also learned in Section 7.2.1 that U.S. regulations require mobile operators to offer emergency call services, which can locate the mobile device to prescribed levels of accuracy. We may not actively use such emergency services on a daily basis (one hopes not), but we still benefit from their around-the-clock availability.

Next, we categorize the many LBSs that are available today. Some LBSs cut across several of these categories, and other LBS may not fit into any of the listed categories. Nonetheless, the following seven categories represent our best attempt to group LBSs into a reasonably small set of categories:

Emergency services: As described above, this is probably the most prevalent LBSs because it is available to all mobile users (in most countries) all of the time, and its performance characteristics are usually regulated by national legislation. In addition to basic 911 services (112 in Europe), some companies offer additional services that should be included in this category. For example, OnStar offers other vehicle-related emergency services, such as automatic crash response. Many of the OnStar services can now be accessed via a smartphone application [27].

Tracking/management: One of the first and most successful commercial LBS areas has been asset tracking and fleet management. For example, in 2013 OmniTRACS already celebrated its twenty-fifth anniversary of providing fleet management services [28]. Today there are many companies worldwide offering similar fleet management or asset-tracking services, many of which are smartphone-based. Note that traditionally fleet management has been largely vehicle-based, but increasingly there is a demand to integrate vehicle and other asset-tracking applications with smartphone platforms. This can range from the smartphone simply being a user interface for the fleet/asset management system to tracking employees based on their smartphones' position.

Map/navigation: Navigation or route/map guidance is another long-standing LBS that has traditionally been implemented on a dedicated navigation device, such as a car navigator. Increasingly, however, people are using their smartphones for such purposes. Google and other companies such as *Here* are offering free smartphone map and navigation applications that can be used in either vehicle or pedestrian mode.

Social LBSs: There are primarily three types of social LBSs, also known as current position-based social networks (LBSNs) or geosocial networks. First are those like *FourSquare*, Google *Latitude*, or *Find My Friends* that have been developed exclusively current position-centric features. There is a great variety within this group. Some are focused on small social groups, such as *Family Tracker* (we might also place this application in the tracking/management category), whereas others are focused on helping users to meet or interact with strangers who visit the same current positions. We also note that "social" features are so ubiquitous within LBSs that in some cases this category blends into the other categories. For example, *Waze* is a vehicle navigation application that adds a social layer to driving. Thus, it could be considered a social LBS, as well as a navigation application. The second type of social LBSs are general social networks, like Facebook, Twitter, LinkedIn, or Instagram, which have relatively recently added location-based features like "check-ins" and geotagging a photo or other post (i.e., status update or tweet). In the second group, many service providers give an option to turn off location-based features for those users who do not wish to share their current position, but it is clear that a large percentage of social network users are utilizing such features. Last, a third type of social LBS includes those applications that aim to aggregate geotagged information from one or more different social networks. For example, one social LBS concept is to display a feed of all posts geotagged within a specified radius of the user's current position.

Information services: Many LBSs offer various information services based on a user's position. Some like *Yelp*, *Google Places*, and *Urbanspoon* are general local business or point-of-interest directories, whereas others target a specific type of business, such as restaurants or movie theaters (e.g., for restaurants *Open Table,* and for movie theaters *Fandango*). In any case, the main idea in this category of LBSs is to filter or customize the information presented to the user based on his or her current position or position history.

Gaming/entertainment: Some LBSs aim to provide games or other forms of entertainment that utilize the users' position. For example, *SCVNGR* is a smartphone-based game, where users perform different kinds of challenges at particular locations. Geocaching is another popular activity, where users try to find "caches" (small boxes containing different objects) hidden at specific geographic coordinates. When a person finds the "cache," he or she can take an object or leave an object in the cache, as well as a small message. There are many smartphone applications that help facilitate geocaching.

Advertising/promotion: Many LBSs aim to deliver advertisements or other kinds of promotional materials that are targeted based on the user's position. In some cases, this is integrated inside a multipurpose LBS, such as a social LBS. In other cases, promotions are delivered via a special-purpose smartphone application, usually in the form of some kind of product discount or "freebie." Examples of such applications include Groupon, Shopkick, and VoucherCloud.

7.4 IMPORTANT LBS CONCEPTS

Despite the diversity of LBSs currently in existence, we can identify a number of concepts that are common to many different LBSs. Section 7.4 describes the most important of these concepts, LBS data types and LBS functions, in brief.

7.4.1 LBS Data Types

First we describe some commonly used LBS data types. The goal of this section is not to be exhaustive but rather to cover those data types that are found in many different LBSs. Also, we note that there is no universal agreement on the definitions of these data types, so we offer our own definitions as used in this book. Figure 7.3 helps to visualize these LBS data types.

Location: A *location* is a specific point in space. Typically in LBSs, we work in two-dimensional space because we are mainly interested in points on the surface of the Earth. Thus, a location can normally be described by a two-dimensional coordinate, such as latitude and longitude. In some cases, a location may be above or below the surface of the Earth, and thus an altitude is also specified. In the case of indoor locations, the altitude may be specified as a floor level, since this is more natural for people to interpret.

Although colloquially we may use the word "location" to refer to imprecise geographical areas, such as a city or a building, it is important to define locations precisely in LBS. The appropriate data type for defining larger areas (i.e., area of interest) will be discussed further below.

Position: A *position* is a current, past, or future location where a mobile device is, was, or will be located. Thus, all positions are locations, but we reserve the term position to refer to the location of a mobile device as a function of time.

Point of interest: A *point of interest* (POI) is a specific location that has a special significance to the user, a group of users, or to the public in general. Thus, a POI is normally described semantically, such as "home," "work," or "Acme Bank." The metadata associated with a POI should include its location. Although the name POI implies a distinct point in space, in most cases POIs actually represent two-dimensional areas (or three-dimensional volumes if one considers, for example, multifloor buildings). Thus, the location associated with a POI is merely a reference location. This reference location should, however, be located within the area or volume covering the POI. Significant places beyond a certain geographic size (e.g., a few hundred meters in either horizontal dimension) may be more appropriate to define as areas of interest (AOIs), rather than as POIs, in order to avoid confusion related to this terminology.

Area of interest: An AOI is a specific geographic area that has a special significance to the user, a group of users, or to the public in general. It is typically defined by a two-dimensional shape, such as a circle or polygon. In cases where polygons are used, then typically the coordinates of each vertex of the polygon are given. Where circles are used, often the center of the circle and its radius are specified. One example of an AOI is a particular city, such as Boston, Massachusetts. Such an AOI may be defined based on the actual city boundaries (i.e., a polygon), or it may be approximated with a much simpler shape (e.g., a circle or ellipse).

In some cases, simple bounding boxes (rectangles) are used to define AOIs. For example, to restrict a search of POIs to within a small AOI, a LBS may use a bounding box defined by its northwest and southeast vertices. Such AOIs may be of only temporary interest to the user, or they may even be completely transparent to the user.

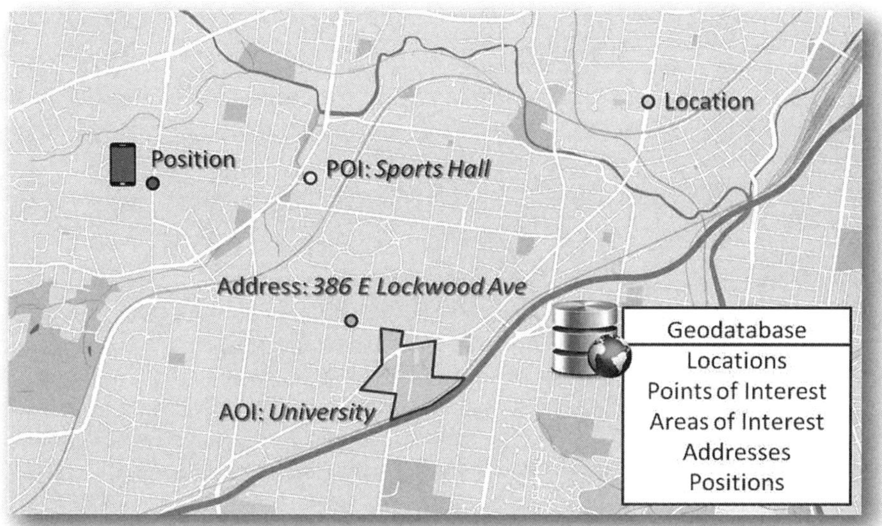

Figure 7.3 Depiction of common LBS data types displayed on a map. Note that many of these data types fall under the broad category *location*, but within this category, there are subclasses of *location*, such as *position*, POI, and *address*. An *AOI* can be represented as a connected set of locations defining the AOI boundaries (e.g. a polygon).

Address: An address is a widely used way to specify a location that provides an alternative to geographic coordinates (i.e., latitude or longitude). Most people do not work with geographic coordinates in daily life, so it is more intuitive to work with an address. Since addresses have been in use for hundreds of years (if not more) and most POIs have a defined address, they are a very convenient way to encode locations. In most countries, an address is composed of a street name (or other named thoroughfare, such as a walkway), an address "number" (which may be a combination of numbers and letters), a city or town, and a postal code. In some cases, only an intersection may be given.

Although addresses are very convenient for humans to work with, there are a few disadvantages when it comes to geospatial computing. For example, one cannot directly calculate distances or relative directions between two given addresses. Thus, an address should also contain metadata that includes its geographic coordinates. Another challenge is that the exact location of an address may be ambiguous or inconsistently defined. Typically an address is associated with a building, which constitutes an area, not a discrete point, which can be a source of this ambiguity. This is the same ambiguity occurring with POIs that cover relatively large areas. An address can also be associated with an AOI, which helps to solve this ambiguity, but this is rarely done in current practice.

Geodatabase: A *geodatabase* (or *spatial database*) is a database optimized to store, query, and retrieve geospatial information, including all of the data types described above. Geodatabases are covered in greater detail in Chapter 5. The most important aspect of geodatabases is that they are capable of creating and using *spatial indices*. A spatial index differs from a normal database index, in that it is designed to efficiently facilitate spatial queries, such as determining the distance between two locations.

7.4.2 LBS Functions

Geofencing: *Geofencing* is the process of monitoring whether a user or device has crossed a specified boundary. The boundary can be defined by a geometric shape, such as a polygon or circle, or in some cases by a line segment. Most commonly geofencing is used for asset tracking and fleet management purposes (e.g., an application that logs every time a truck enters or leaves the loading dock area). It could also be used for safety and security applications, for example, to alert monitoring personnel if a monitored person (e.g., a dementia patient) has left a controlled area. Another commonly referenced application of geofencing is for LBS that incorporates advertising or promotions into its functionality—it may be desirable to track whether a user crosses within a specified distance of a particular business and push a promotional ad to the user's phone (e.g., "Get a free bagel with any purchase at Bob's Coffee, only 100 meters away!").

Some commercial services exist to automatically perform geofencing operations in the cloud. The typical architecture is that positioning is performed continuously at a predefined interval in the background on the user's device and then sent to a cloud server. Geofencing calculations are performed in the cloud and the LBS's server application initiates a push operation based on the user's current position and preferences. Two such examples of cloud-based geofencing are Qualcomm's Gimbal™ platform and Google Play Services (the latter for Android devices only).

Geocoding: *Geocoding* is the process of determining the geographic coordinates (i.e., location) associated with another geographic entity, such as a POI, an address, or a city. If the mobile device contains this information in a database, this is achieved with a simple table lookup. It does not even require the use of a geodatabase, as geographic coordinates can easily be stored and retrieved in a traditional relational database. Due to the storage limitations in a mobile device, however, geocoding is most commonly implemented as a cloud-based service. Modern mobile operating systems have built this functionality into their developer APIs, especially for the case of geocoding addresses.

Geocoding can also refer to various techniques to interpolate geographic coordinates of addresses from a road network, where the only known information is the road segments and their address ranges (e.g., 100 block of Main Street). In most cases, however, this process has already been performed and verified by, for example, the company providing the mobile operating system and its affiliates, so application developers rarely need to understand the details. It is important to note, however, that geocoding can introduce errors, due to the fact that this interpolation process is imperfect, in addition to the ambiguities with addresses and POIs described above.

Reverse geocoding: As the name implies, *reverse geocoding* is essentially the same process as for geocoding but in reverse. Given a set of geographic coordinates (or location), determine the nearest address or POI. In the case of AOIs, this can amount to determining whether a specific location is within any defined AOI. Depending on the application, it may be enough to determine that a location is within a particular city or postal code, but most often the application requires a complete address. Again, most mobile operating systems provide this capability through their developer APIs, but the same caution regarding possible errors applies. Especially in places where addresses are very dense, errors in the geographic coordinates may hinder the identification of the correct or desired address.

In some implementations, it may be possible to specify a bounding box or a radius as an input to the reverse geocoding function, as opposed to merely a discrete point in space. This amounts to defining an AOI, and the reverse geocoding process then becomes that of determining a list of addresses or POIs located within the defined AOI.

"Find nearest POI": A common LBS function (for which there is no standard name) is to find the POI with specified properties nearest to a reference location (e.g., the restaurant closest to the user's current position). This is similar to the case of reverse geocoding, where the AOI of interest is defined as a circle with the center being the reference location. The difference is that the goal of this function is to find the *nearest* POI, which means that the function should not just search inside a circle of defined radius, but it should continue searching in an expanding radial fashion until it encounters at least one POI matching the specified properties (i.e., the set of all restaurants).

Actual implementations of this function may differ in the search strategy, but this is certainly a function that benefits from the efficient geometric operations that can be performed on a spatial index. Last, we note that the term *nearest* is used in an abstract sense. Methods other than the geometric distance may be used as a measure of "nearness." For example, in many LBS navigation applications, it is more appropriate to use driving time as the distance measure. In this case, the application must be able to perform some form of *pathfinding* (see below).

Pathfinding: *Pathfinding* is the process of determining a path (or route) between two specified locations, subject to a set of constraints and optimized according to a specified criterion. The most common example in LBS is finding a driving route in a road network. The road network provides the set of constraints (e.g., the route should follow the roads and travel at specified maximum velocities) and the typical criteria of optimality are driving distance or driving time, which should be minimized.

The classical algorithm to find an optimal path in a prescribed graph (e.g., road network) is Dijkstra's algorithm (published in 1959) [29]. As the number of edges (e.g., intersections) and vertices (road segments) in the graph increases, the running time for this algorithm also increases (linearly with the number of edges and linearithmic with the number of vertices). For large road networks, more efficient algorithms are needed. One such commonly used algorithm for pathfinding is called A^*, which uses a heuristic function to guide the path search in the direction of the destination [30]. Even A*, however, may not be fast enough for very large road networks. Various approximation algorithms and graph preprocessing techniques have been devised to compute optimal or approximately optimal paths in less running time. Further challenges arise if the pathfinding algorithm must take into account time-dependent factors, such as traffic congestion. This is an active area of computer science research—for recent reviews, the interested reader is referred to [31] or [32].

7.5 CURRENT LBS MARKETPLACE

In this section we will take a look at the recent situation in the marketplace concerning LBSs. This can be viewed in two different ways—macroscopically (e.g., market size and trends) and microscopically (user desires and attitudes toward LBSs). For someone interested in developing LBSs or otherwise working in this industry, both angles are important to understand.

7.5.1 LBS Market Size and Overall Trends

It is difficult to gauge the exact size of the LBS market, but it is certainly clear that the market is experiencing heavy growth. The opportunities for application developers and service providers to introduce new offerings to the market are many. There are already hundreds of thousands of LBS applications existing in the major app stores today, but new ones are being added every day.

Various market studies have been carried out to assess the current market revenues from LBSs and forecast future growth. The findings of these studies vary greatly, in large part because the organizations conducting the studies have varying definitions of what revenues to include under the category of LBSs. For example, smartphone and tablet revenues were greater than $300 billion globally in 2013. Since phone and tablet manufacturers include free LBSs with their products, one must consider how much of these revenues to attribute to LBSs. For a company like Google, whose revenues are primarily generated from advertisements, we can consider the revenues generated from mobile ads. These exceeded $8 billion in 2013. At least some of these revenues can be attributed to the many LBSs that Google offers to users free of charge. Even the most conservative estimates place LBS revenues at over $1 billion in the United States and Europe alone (2012 estimate). Another market analysis firm, Gartner, forecasts that revenues from consumer LBSs will reach $13.5 billion in 2015.

One common trend in the mobile industry, which may be attractive for some application developers, is for successful startups to be acquired by larger companies, such as Google, Apple, and Facebook. Since the cost of the larger acquisitions must be reported for regulatory reasons, this offers another way to gauge the value of LBSs in the market. For example, the successful social road navigation start-up *Waze* was acquired by Google in June 2013 for $966 million. There are many other such examples in the LBS industry, although Waze is perhaps the largest LBS acquisition in recent years.

In summary, there is a great deal of activity and growth in the current LBS market, which makes it an exciting industry in which to work and study. At the same time, it is a rapidly changing industry, so it important to update one's knowledge of the current trends frequently.

7.5.2 User Behavior and Attitudes Toward LBS

In this section, we will examine how users currently use LBSs and what perceptions and attitudes they have concerning services currently in the market. One way to gauge users' desires and attitudes is through user surveys. Several studies of this nature have been carried out in recent years, attempting to understand LBS users' experiences and perceptions. In particular, we will focus on how users would like existing services to develop, and what new LBSs should be developed in the future.

The largest LBS user research conducted in recent years was that of the Pew Research Center, which carried out three surveys between May 2011 and May 2013. In the most recent survey, it found that 74% of adult smartphone owners use LBSs, such as getting directions or other information based on their current position. An increasing number of users are using location in their social media activities (e.g., tagging their posts with a location). The survey reports that this percentage increased from 14% in 2011 to 30% in 2013. A relatively small percentage of those surveyed are using "check in" services, such as those provided by Foursquare, Facebook, or Google Plus. This figure varied between 12% and 18% over the course of the three surveys [33].

Another survey conducted in 2013 by the Finnish Geodetic Institute (FGI) found that about 27% of people surveyed used a mobile phone or tablet every day for positioning purposes. An additional 22% used such devices for positioning almost every day, and nearly 25% indicated that they used such devices for positioning "often" (defined as a few times per week). The FGI survey also examined the purposes for which people are currently using LBSs and found that the most common purposes were car navigation (82%), pedestrian navigation (68%), "sports and leisure" purposes (65%), and public transportation purposes, such as planning journeys (51%) [34].

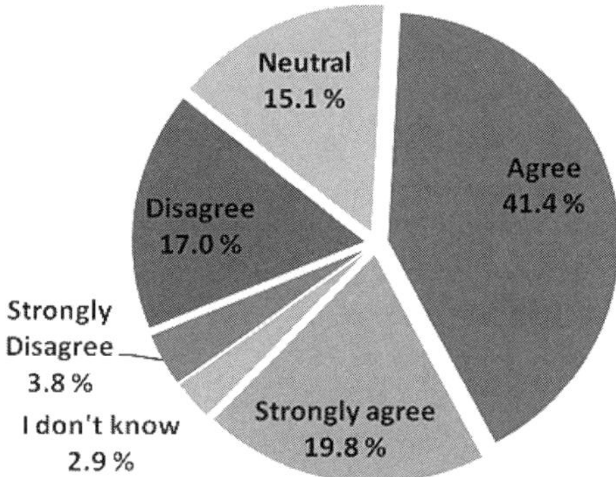

Figure 7.4 Result of survey where people were asked about how positioning accuracy affects their usage of LBS [34].

Last, the FGI survey aimed to determine how respondents' usage of LBS would be affected if the positioning accuracy were to improve significantly. Respondents were asked to indicate whether they agreed with the following statement:

If the accuracy of positioning in a low-cost device would improve from current levels (5-10 meters) to 0.5 meters, my usage of positioning would increase significantly.

The responses to this question are summarized in Figure 7.4.

Another study reports on a survey of students at two European universities, which sought to examine the students' attitudes toward LBS [35]. The authors identified 10 different LBS categories and asked survey respondents to rate these categories in terms of their willingness to pay (e.g., monthly subscription) for an LBS belonging to each of these categories. The top three categories in terms of willingness to pay were 1) emergency-related applications (e.g., getting faster help in the case of a road accident), 2) navigation-related applications (e.g., multimedia navigational guide for tourism), and 3) well-being–related applications (e.g., health and fitness advice based on one's mobility patterns). The bottom three categories (ordered from highest to lowest willingness to pay) were 1) geotagging (e.g., ability to add location information to photos), 2) entertainment (e.g., games tailored to one's current position), and 3) advertising (e.g., receiving product offers based on one's current position. These give a sense of what LBS categories users are willing to pay for directly through subscription fees.

Another interesting factor examined in this survey was the willingness of users to accept that a LBS would consume additional power, in order to locate the user in a continuous manner. The results found that about 30% of respondents would accept up to 1% extra power consumption for this capability and that about 42% would accept up to 5% extra power consumption. Only a small number (about 14%) would accept higher levels of power consumption, and an equally small number would not allow any extra power consumption for this capability. This last group we can interpret as not willing to "spend" any of their battery power budget on continuous location tracking. Overall, we can conclude that most users are only satisfied if an LBS they are using consumes a relatively small amount of battery power (i.e., if we consider a power budget, less than 5% should be expended on continuous location tracking).

Finally, this survey examined privacy preferences of LBS users concerning sharing their location information (specifically whether they consent to their mobile operator sharing this information). Less than 1% of respondents had no privacy concerns whatsoever concerning usage of their location information. A sizeable number (about 20%) responded that they would not give consent for their mobile operator to use their location information under any circumstances (including emergencies). The largest group of respondents (about 39%) indicated that they would give consent if the operator asked each time before distributing their location information to a third party. These results indicate that the majority of LBS users want to retain tight control over who has access to information about their location (despite the fact that many current LBS providers may not operate in this way).

7.6 SUMMARY

In this chapter we examined the popular topic of LBSs. We began by formally defining a LBS as "any application or service that exploits the position of a mobile device, where the position is determined through automatic means." We then reviewed how LBSs came into being, paying particular attention to the convergence of three key technologies: mobile, Internet, and location (i.e., positioning) technologies. We also highlighted the early commercial actors in this history.

Next, we discussed the different types of LBSs in existence currently, grouping them into seven categories: emergency services, tracking/management, map/navigation, social LBS, information services, gaming/entertainment, and advertising/promotion. Section 7.4 covers important LBS concepts, such as common LBS data types and functions. Last, Section 7.5 looks at the LBS market. This includes discussion on the size and growth of the LBS market and the current trends, as well as user behaviors and attitudes.

In conclusion, the topic of LBSs is diverse but fascinating in many different ways. There are many services in which location plays a key role, and the availability of this information in smartphone platforms opens up vast new opportunities for application developers and service providers. Indeed, LBS is changing the mobile industry in dramatic ways and driving future growth. It is not a stretch to conclude that LBSs are changing the way people live and work.

References

[1]   Virrantaus, K., et. al., "Developing GIS-supported location-based services," *Proc. of WGIS'2001–First International Workshop on Web Geographical Information Systems,* Kyoto, Japan. 423–432, 2001.

[2]   Open Geospatial Consortium Inc., 2008, OpenGIS Location Services (OpenLS): Core Services, Available from http://portal.opengeospatial.org/files/?artifact_id=22122.

[3]   Foursquare, *About Foursquare*, Available from https://foursquare.com/about.

[4]   DeFeo, J. E., "The future mobility market," *Cellular Business*, Jan 1993, pp 24-36.

[5]   Google, Inc., Google Books Ngrams Viewer, Available from https://books.google.com/ngrams.

[6]   Harter, A., and A. Hopper, "A distributed location system for the active office," *Network, IEEE 8*, no. 1, pp. 62-70, 1994.

[7]   Asthana, A., M. Cravatts, and P. Krzyzanowski, "An indoor wireless system for personalized shopping assistance," *Mobile Computing Systems and Applications, 1994. Proc., Workshop on*, pp. 69-74. IEEE, 1994.

[8]   The World Bank Group, Data | The World Bank, Available from http://data.worldbank.org/, 2013.

[9]   Cailliau, R., *A Little History of the World Wide Web*, Available from http://www.w3.org/History.html, 1995.

[10] US Naval Observatory, 2013, GPS INFO, Naval Oceanography Portal, Available from http://www.usno.navy.mil/USNO/time/gps/gps-info.

[11] Zak, Antoly, 2013, *GLONASS Network,* Available from http://www.russianspaceweb.com/uragan.html.

[12] Manuta, L., "Intelligent vehicles, smart satellites," *Satellite Communications*, Feb. 1992, pp. 21-24.

[13] RDSS, LLC, 2013, *RDSS.COM*, Available from http://www.rdss.com/.

[14] Federal Communications Commision, 1996, *Revision of the Commission's Rules to Ensure Compatibility with Enhanced 911 Emergency Calling Systems,* Available from http://transition.fcc.gov/Bureaus/Wireless/Orders/1996/fcc96264.txt.

[15] Federal Communications Commision, 2000, *Revision of the Commission's Rules to Ensure Compatibility with Enhanced 911 Emergency Calling Systems,* Available from http://transition.fcc.gov/Bureaus/Common_Carrier/Orders/2000/fcc00326.pdf.

[16] Wortham, L. C., "Vehicle Locating and Communicating Method and Apparatus," U.S. Patent 5,155,689, filed Jan 17, 1991, and issued Oct. 13, 1992.

[17] Barnum, A., "High-Tech Tracker Seeks And Finds Stolen Vehicles," *Chicago Tribune*, April 12, 1992.

[18] Barabba, V., et al., "A multimethod approach for creating new business models: The General Motors OnStar project," *Interfaces 32,* no. 1, pp. 20-34, 2002.

[19] Bouve, L. W., W. T., Semple, and S. W. Oxman, "System and methods for remotely accessing a selected group of items of interest from a database," U.S. Patent 6,385,622 B2, filed Jan 11, 1995, and issued May 7, 2002.

[20] Chen, M., and I. S. "Small, Time and location based computing," U.S. Patent 5,642,303, filed May 5, 1995, and issued June 24, 1997.

[21] INS-NEWS, Press Release: "GSM Positioning System Enables Instant Vehicle Tracking," 4 March, 1998, Available online: http://bit.ly/11mU5Uu.

[22] Cision, Inc., Press Release: "Shareholders Upbeat at CellPoint's Annual Meeting," December 1, 1999, available online: http://bit.ly/12mbtKV.

[23] Combee, B., et al., *Palm OS Web Application Developer's Guide*, Rockland, MA: Syngress, 2001.

[24] Business Wire, Press Release: "3Com Delivers the Palm VII Organizer for Out-of-the-Box Wireless Internet Access," 24 May 1999, available online: http://bit.ly/12m0aT1.

[25] Amor, D., *Internet Future Strategies: How Pervasive Computing Services Will Change the World*, Upper Saddle River, New Jersey: Prentice Hall, 2002.

[26] Unni, R., and R. Harmon, "Perceived effectiveness of push vs. pull mobile location-based advertising," *Journal of Interactive Advertising 7*, no. 2 (2007): 28-40.

[27] OnStar, LLC, 2013, *OnStar RemoteLink mobile app: control your vehicle from anywhere,* Available from https://www.onstar.com/web/portal/connectionsexplore?tab=2&g=1.

[28] Gomes, A., 2013, *Omnitracs Celebrates 25 Years of Accomplishments and a Successful Future,* Available from http://www.omnitracs.com/media/blog/tags/omnitracs-25th-anniversary

[29] Dijkstra, E. W., "A note on two problems in connexion with graphs," *Numerische mathematik 1*, no. 1 (1959): 269-271.

[30] Hart, P. E., N. J. Nilsson, and B. Raphael, "A formal basis for the heuristic determination of minimum cost paths," *Systems Science and Cybernetics, IEEE Transactions on 4*, no. 2 (1968): 100-107.

[31] Laporte, G., "Fifty years of vehicle routing," *Transportation Science 43*, no. 4 (2009): 408-416.

[32] Pillac, V., et al., "A review of dynamic vehicle routing problems," *European Journal of Operational Research*, Vol 255, 2013, pp. 1-11.

[33] Zickuhr, K., "Location-Based Services," Pew Research Center, Available from http://bit.ly/1eDWh3v, Sept. 2013.

[34] Finnish Geodetic Institute, "End-users and user applications for the public precise positioning service based on the national GNSS network," (Internal research report), November 2013.

[35] Lohan, E., et al., "End-user attitudes towards location-based services and future mobile wireless devices: The students' perspective," *Information 2*, no. 3 (2011): 426-454.

Chapter 8

Context Awareness

Up to this point, this book has focused on two fundamental aspects of geospatial computing: (1) methods to compute one's position and (2) methods to store and visualize location-referenced information. These elements provide the basis for *location awareness* and for creating and consuming mobile LBSs, a topic that was covered in detail in Chapter 7.

The next topic, *context awareness*, may at first seem like a departure from the topics covered in the previous chapters, which focused mainly on *location*. In fact, this chapter is a broadening of our focus because, as we shall see, location awareness falls under the larger umbrella of *context awareness*. We will consider various definitions of this term, but a precursory (and open-ended) definition might simply be that context awareness is the superset of location awareness and many other types of awareness about a mobile user's environment, activities, and state of being.

Although context awareness has existed as a computer science research topic for nearly 20 years, the topic has in recent years received increasing interest, largely due to the proliferation of mobile devices in everyday life, as discussed in Chapter 1. It is simultaneously an extremely promising yet challenging topic, given the wide range of capabilities it encompasses.

This chapter presents context awareness at a conceptual level and provides historical perspective on the development of the field. In addition, we attempt to show at a practical level how different aspects of contextual awareness can be implemented in a mobile device. Due to space limitations, we have selected one popular mobile platform, the Android operating system, in order to show specific code examples of several concepts. We will also point readers toward methods of the Android API classes that they can use to implement their own context-aware applications. Similar capabilities are available in other major mobile platforms, such as iOS and Windows Phone 7/8, but unfortunately space does not allow us to demonstrate them here.

8.1 DEFINING CONTEXT AND CONTEXT AWARENESS

Context is one of those words, like system and information, that is difficult to define precisely. It is used prominently in many fields, including linguistics, psychology, neuroscience, law, and computer science, but ironically the definition of context depends heavily on its context of use. Loosely speaking, we can say that context is a category of information, but there is no general agreement on which types of information belong to this category.

Some researchers have studied the etymology of the word context and have even attempted to formalize and to build consensus concerning the definition [1–4]. Although these semantic topics are beyond the scope of this chapter, we require some working notional definition of context before we can define and examine context awareness (this chapter) or discuss *contextual reasoning* (Chapter 9).

Note that whatever definitions we adopt, they will surely reflect the particular focus of this book, namely mobile geospatial computing. In fact, in defining these concepts, we notice a definite trade-off between generality and concreteness. We have tried to be as general as possible without losing our focus on how context awareness can be implemented in mobile devices.

8.1.1 What Is Context?

In the *Merriam Webster Dictionary* [5], we find two definitions of the word context:

- The parts of a discourse that surround a word or passage and can throw light on its meaning;
- The interrelated conditions in which something exists or occurs: ENVIRONMENT, SETTING.

In this book, we adopt the second definition because we are not directly concerned with human discourse but rather with conditions of an environment or setting (i.e., geospatial information) that can be sensed by machines (i.e., computers and sensors). Clearly, these two definitions are interrelated in the sense that discourse can be (and most usually is) used as a representation of an environment or setting.

In other words, natural language is a common form in which contextual information is encoded. Our focus though is on techniques to sense and represent context automatically. Hence, when we refer to *context*, we refer directly to the conditions in the environment. When we use the term *contextual information*, we refer to representations of context, either in natural language or other form. Furthermore, when we want to explicitly distinguish between the two types of context defined above, we will use the terms *discourse context* and *conditions context*, respectively.[17]

In lieu of a formal and generally accepted definition of context, the above dictionary definition (second one) is broad enough to serve as a working definition for use in exploring context awareness. We will further elucidate this concept with a framework and example below.

There are a few general notes regarding context to cover before we dive deeper into the topic. First, note that our definition implies that the specific "conditions" relevant to context are dependent on the "something" to which the context applies. In other words, context is always specified from a particular viewpoint or *frame of reference* of an object or person. This is similar to the concept in physics, where a physical measurement like velocity is defined within a particular frame of reference. Within the domain of mobile geospatial computing, our frame of reference most often will be the mobile device itself, although oftentimes we take the frame of reference to be the same as that of the mobile user.

Second, we must choose some techniques for describing a particular context that help to build a framework for our contextual information. If our goal is to describe a particular context in natural language, then we might employ the classic technique of journalism (since journalism is an age-old craft for describing conditions and events), known as the *five Ws*: who, what, where, when, and why [6]. In fact, this technique dates back at least to the late second century BC when Hermagoras of Temnos defined seven elements of circumstance, which include (in addition to the five Ws) "in what manner" and "by what means" [7].

Using these questions as a starting point (with a slightly different order), we list possible elements of a particular context with a demonstrative example:

What: A small, impromptu gathering of colleagues.
Who: Mary, a smartphone user, as well as three of Mary's coworkers who are nearby.
Where: 60.1609°N, 24.5460°E (WGS84); inside the main lobby of the FGI, specifically inside Mary's pocket.
When: Friday, 20 April 2014 at 12:03 p.m.
Why: This gathering occurred because Mary and her colleagues are going out to lunch together. They are waiting for a fifth colleague, Steve, to arrive.

---

[17] The word context is also used in some image analysis literature to refer to pixels in the vicinity of a particular pixel (e.g., the neighbors of that pixel). See references [8–10] for examples. This choice of terms is, in fact, related to the *discourse* definition of context; just as the surrounding discourse can shed light on a particular word or phrase, the surroundings of a particular pixel can help in interpreting its value. In any case, this meaning of context is in some ways different from conditions context, and we mention this usage merely to caution the reader not to confuse the two.

In what manner: Mary's smartphone is experiencing small, sporadic movements, consistent with the phone being in the pocket of someone who is standing and having a casual conversation.

By what means: All of the above information has been sensed or reasoned by the sensors and software existing in a smartphone, or acquired via a networked resource. In this case, the smartphone is a Samsung Galaxy S4 with Android 4.3 OS, which includes a GNSS receiver, WLAN-based positioning engine, Bluetooth module, microphone and audio analyzer, ambient light sensor, accelerometers, gyroscopes, and magnetometers.

This account of the situation is not likely to yield a Pulitzer prize in journalism, but it is enough to get a basic sense about what the mobile user is doing, as viewed from the perspective of her mobile device. We will see later how this contextual information can be used to create context-aware applications that record a user's activities and even anticipate his or her needs.

There are of course many other contextual elements we could list, including details about the smartphone itself (e.g., battery level and mobile network that it is connected to) or about the immediate surrounding environment (e.g., the weather conditions outside). Like a good journalist, however, we only include the most relevant facts that are necessary to understand the situation at hand. If, for example, the weather outside were particularly noteworthy, we might have included it in the list, but we can assume from its omission that most probably it is nothing exceptional or affecting to the situation.

Last, we note that there are some elements omitted from this list that are certainly both contextual and relevant, such as details about the user's state of mind or what the topic of conversation is. This illustrates the point that our contextual information will in some ways always be incomplete. In the domain of mobile geospatial computing, we are limited to the information that can be obtained or reasoned by a mobile device. As mobile technology rapidly advances, the breadth of information available to mobile devices will surely increase, but it will always face limitations, especially compared to the robust and subtle sensing ability of the human being.

8.1.2 What Is Context Awareness?

Now that we have a basic notion of context and a simple example to aid our understanding, we can define the term *context awareness*. Although this term inherits the same difficulties facing the term context, it is otherwise straight forward to define. Context awareness is the quality of having knowledge (being aware) of context. We say a device or application is *context-aware* if it possesses this quality, and we define *context-aware computing* as the set of computational techniques designed to obtain and use context.

Using the journalistic framework described above, we can describe one objective of context-aware computing as the building up of a description of a particular situation or context. As an ultimate goal, this description should be indistinguishable from the description that a perceptive (human) journalist would compose. Thus, the process can be roughly framed as 1) acquiring information from sensors and other sources and 2) translating this information into prose (or ordinary spoken language).

Just as a person can have different levels of awareness, a device or application can have different levels of context awareness. To a certain extent, nearly every mobile device is context-aware in the sense that it knows certain details about its state (e.g. , battery level, networks it is connected to, and whether or not a phone call is taking place with it) and perhaps about its user or owner (name, phone number, etc). We arbitrarily distinguish, however, between context-aware and non-context-aware devices based on the amount and breadth of contextual information that the device can obtain and whether it possesses sensors and software that are employed specifically for the purpose of obtaining contextual information. Admittedly, this is a blurry definition, and this type of classification is often a matter of opinion.

Don't be surprised if some of this starts to sound a bit "sci-fi." The goal of creating a context-aware device is related to one of the primary prerequisites of artificial intelligence (AI). A machine exhibiting AI must have a general representation of the world, especially concerning the objects and beings in its immediate surroundings [11]. We shall see, however, that a device can achieve some elements of context awareness without even coming close to achieving AI. Hence, the two subjects are related but distinct.

Last, in order to aid the reader's understanding, we present briefly an example of a real-world, present-day context awareness product—Google Now. The web site for this product provides a succinct description of its context awareness capabilities [12]:

Google Now gets you just the right information at just the right time.

It tells you today's weather before you start your day, how much traffic to expect before you leave for work, when the next train will arrive as you're standing on the platform, or your favorite team's score while they're playing. And the best part? All of this happens automatically. Cards appear throughout the day at the moment you need them.

Of course, none of these capabilities would be possible without a source of contextual information, such as the user's position, preferences, or activity (e.g., leaving for work). It is precisely the acquisition of this type of information that we will explore further in this chapter; we will learn how to process it in Chapter 9.

8.1.2.1 Is Location Awareness Part of Context Awareness?

As stated above, location awareness is one element of context awareness, thus we could claim that any device that is location-aware is therefore, to a certain extent, context-aware. In fact, location awareness can be viewed as an important pioneer and forerunner in context awareness. There is no doubt that location is one of the most important elements of context, hence the focus of much of this book on positioning methods. This is one aspect of the "geospatial era," discussed in Chapter 1, where many forms of computing are adopting location as a central contextual element. The most recent advances in mobile technology, however, point to a future where mobile devices will be aware of much more than just location. In fact, positioning technologies are becoming so ubiquitous that a location-aware device may not seem noteworthy enough to merit the special distinction of being context-aware. Indeed, our view is that location awareness is a necessary precursor to context awareness (at least for the vast majority of applications) but does not in itself imply context awareness.

8.1.2.2 Relation Between Context Awareness and Situation Awareness

Some readers may note that a related term *situation awareness* (or *situational awareness*) is commonly used in certain disciplines. The distinction between context awareness and situation awareness is largely of a historical nature rather than conceptual. The former comes from mobile computing researchers, whereas the latter is rooted in military aviation.

Originally, situation awareness referred to the pilot's "awareness of conditions and threats in the immediate surroundings" [13], and its importance was known as early as World War I, especially by German flying ace Oswald Boelke [14]. Eventually situation awareness was also used to refer to machine awareness of such conditions and threats. Furthermore, the term became popular in a plethora of fields outside of aviation applications, including maritime[18] or ground-based operations, and even outside of military applications [15]. Hence, today the concepts are essentially the same, and the two terms can be used almost interchangeably. Because context awareness is more commonly used in mobile computing, whereas situation awareness is more commonly used in military, safety, and security applications, we will primarily use the former in this book.

8.2 WHY IS CONTEXT AWARENESS IMPORTANT?

Now that we have a basic understanding of the concept "context awareness," it is worth asking the question, "Why is context awareness important?" Given its

---

[18] Also known as maritime domain awareness.

prominence as a topic in computer science research, there should be compelling reasons for why it is studied.

As Anind Dey, one of the pioneers of context awareness, and others have pointed out, the goal of creating context-aware devices is largely about improving the richness and ease of human-computer interaction [16]. If a computer can be made more aware of the conditions and environment in which it and its users reside, then there will be less need for the users to explicitly tell the computer what it is they need the computer to do. To a certain degree, the computer will "know" what our needs are. This is because developers will have prescribed what is typically needed in a given situation into an application designed to serve such needs (or allowed the user to specify his or her needs in given situations).

A simple example could occur when a mobile device detects that you will be late for a meeting (because you are 30 minutes by car from the meeting place with only 15 minutes until the meeting starts), and could automatically send a message to a contact saved in your mobile device and identified in your mobile calendar as the organizer of the meeting. Therefore, to a large extent, context awareness is about increasing the automation capabilities of our computing devices, or *minimizing* the amount of human-computer interaction that is required to perform a given task.

Many other examples of such context-enabled capabilities could be listed here, but like many worthy scientific and technological endeavors, it is impossible to say what the most important reason for pursuing context awareness is. It is a bit like speculating in the early 1960s about the importance of the Internet (or Intergalactic Computer Network, as it was known in an early phase) that we see so clearly from our vantage point in the twenty-first century. Certainly at that time computer scientists could see the value of networking together four to eight mainframe computers to share resources and transfer data, but they could hardly envision at that time how such entities as Google, Facebook, Twitter, and Skype would change the way that billions of people operate on a daily basis; how businesses would operate; and even how revolutions would be organized! Similarly, although we can see some of the immediate promises of context awareness for mobile computing, we must rely on our intuition and curiosity to believe that making computers aware of context is an effort that will pay off in countless, unimaginable ways.

8.3 HISTORY OF CONTEXT-AWARE COMPUTING

It is worthwhile to review the history of context awareness from the literature and see how the concept has developed and been employed. Reviewing this history is challenging, however, due to the wide and varying use of the concept "context," as discussed above. Some argue that the topic of context can be traced back to

Frege [18, 19] or Russell [20], where the role of context in determining the meaning of symbols and words is discussed [17–20].[19] These philosophical issues regarding the word "context" are not, however, essential for a practical understanding of context awareness, so we will only mention them in passing during this brief review.

In a similar vein, we can find studies from the 1950s and 1960s, where context is used in various applications, such as machine recognition of handwriting, printed text, and verbal speech [21–23]. These works, however, are primarily concerned with discourse context rather than conditions context. Discourse context is fundamentally different from conditions context because in the former there is (by definition) a human in the loop, whereas representations of the latter could be generated automatically.[20] As we stated at the onset, we are interested in automatic methods; hence these studies are only of peripheral interest.

In some sense, however, discourse (i.e., natural language) is relevant to context awareness because it can be used to infer conditions context. Thus, *natural language processing* can help provide inputs to context-aware applications. A full survey of natural language processing techniques is beyond the scope of this book, so the reader is referred to Porzel's extensive coverage of the topic in [24].

As mentioned above, the subject of context awareness exhibits strong ties to the field of AI. As early as 1963, one of the "fathers of AI," John McCarthy, described the concept of a *fluent* (later called *situational fluent*), which we might also call a *contextual proposition* or *contextual statement*. A fluent is expressed as a function of a *situation*, which is described as "the complete state of affairs at some instant of time." One simple example of representing a fluent that McCarthy provides is `raining(s)`, which indicates that it is raining in situation s. From this example, we can clearly see the parallel between a set of *fluents* for a particular situation and a context [25].

It is curious to note, however, that McCarthy does not explicitly use the term context until a 1987 paper titled "Generality in Artificial Intelligence" [26]. Even then, he does not provide an explicit definition of context nor comment on its relation to the concepts of fluent or situation. Nonetheless, this paper seems to be the spark that set off an intense effort to study and even formalize the concept of context. By the early 1990s, we can find important works on formalizing context and on contextual reasoning [1, 2, 27, 28]. Although this literature does not

---

[19] Russell does not explicitly use the term "context," but one can recognize its implicit presence.

[20] Computers, of course, can simulate human discourse and can even generate discourse about the conditions context. Still, we feel justified in making this distinction because it seems that a computer capable of human-like discourse requires first some knowledge of at least some elements of conditions context. Hence, computer-generated discourse would constitute a particular use of conditions context, rather than a method of acquiring conditions context.

directly herald an era of context-aware devices, it constitutes important theoretical groundwork in the development of context awareness as a research topic.

Another important precursor to context awareness can be seen in the often cited article "The Computer for the 21st Century," published in *Scientific American* [29]. In this highly readable article, Weiser envisions a future state of technology where context awareness plays a crucial (albeit unspoken) role:

Sal awakens; she smells coffee. A few minutes ago her alarm clock, alerted by her restless rolling before waking, had quietly asked, "Coffee?" and she had mumbled, "Yes."

This fictional account goes on to describe a number of automated and assisted functions that are performed by *ubiquitous computing* devices, which in many cases can be considered to be context-aware. When this article was published in 1991 it probably sounded like mostly science fiction. With the recent advent of wireless devices like Fitbit (which can monitor sleep periods), such scenarios are now on the verge of reality.

For the first reported implementation of a context-aware device, we look to the Active Badge Location System, developed at the Olivetti Research Laboratory in Cambridge, England, in the early 1990s [30]. Not surprisingly, this system was focused exclusively on creating *location awareness*, especially for employees in an office environment. A few years later, Schilit and Theimer were the first to explicitly use the term context-aware computing, which they defined as "the ability of a mobile user's applications to discover and react to changes in the environment they are situated in" [31].

By the mid1990s, the study and development of "context-aware applications" becomes an active area (e.g., PARCTAB, stick-e notes [32], CyberGuide [33], CyberDesk [34]). The reader can find an excellent snapshot of the turn-of-the-century, state-of-the-art in context-aware computing in a special issue of the journal *Human-Computer Interaction*, published in 2001 [35]. What is clear from the literature of this period is that context awareness had become a strong research focus for several research groups, typified by prototype and architecture development. It is also clear, however, that an abundance of challenges still remained to achieving rich and robust context awareness.

During the past 10 years, context awareness research has continued to press forward, and it has begun to work its way into several commercial products. Several frameworks for context awareness have been developed by researchers. One challenge that remains, however, is that no widespread standard yet exists for representing contextual information.

As far as commercial products, the example of Google Now was given above, but there are plenty of other examples as well. Dexetra currently offers an application called "friday," which keeps track of a user's daily activities, such as commuting, activity on social networks, and various uses of the user's phone (calls, photo-taking, etc.), as well as the location where these features were used.

Qualcomm Retail Solutions offers a context awareness platform, called Gimbal, with SDKs available for iOS and Android. Already in 2012, Samsung's Galaxy S3 smartphone was marketed as a context-aware device with its ability to recognize when a face is looking at the screen and control whether the screen stays on base on this information. The Motorol Moto X, released in 2013, also exhibits many context-aware features. These include sensing when the phone is picked up (in order to turn on the screen and display notifications) and sensing certain hand gestures (in order to activate the camera application). An application called Motorola Assist[21] uses context to detect activities like driving and sleeping. It allows users to predetermine how to handle or respond to incoming text messages and phone calls, based on the current activity (e.g., reading text messages to the user while driving).

8.4 EXAMINING CONTEXT IN DETAIL

The remainder of this chapter examines in more detail the various elements of context first introduced in Section 8.1.1. Some of these elements are more straightforward to detect in a mobile device than others. For the more challenging elements, we will provide only precursory coverage in this section.

8.4.1 What: The Activity Context

The first element of context is a high-level description of what activity is taking place or what the overall current situation is.[22] Examples could include: driving a car, waiting at a bus stop, having lunch, and participating in a meeting. In practice, there can be many possible ways to formulate such high-level activity or situation descriptions, as well as many different techniques for classifying and sensing them. The overall objective of classifying and sensing activities is often called *activity recognition*, and a great deal of research has been conducted in this area alone.

The optimal set of techniques to detect high-level context is specific to the set of sensors or other data available, the scope of activities that one desires to detect, as well as the specific requirements of the application (such as whether or not the context must be detected in real time). Therefore, it is very difficult to provide general guidance on how to perform activity recognition. In broad terms, however, we can use one or more techniques from the fields of *machine learning* and

---

[21] At the time of this writing, Motorola Assist was compatible with four different Motorola smartphone models, including the Moto X.

[22] We recognize that "situation" is essentially a synonym for context, but it is used here to embody the highest level of contextual information that functions as a summary of the context. We also note that a context may sometimes be characterized by a great deal of *inactivity*, but even "sitting idly" can be one classification within a set of activity classifications.

pattern recognition. Specifically, the following signal processing techniques have been used for activity recognition:

- Decision tree;
- Hidden Markov model;
- Naïve Bayes classifier;
- Support vector machine;
- Kalman filter;
- Particle filter.

We will provide further details on some of these techniques in Chapter 9. Machine learning and pattern recognition are fields with a rich set of literature, and we can only scratch the surface in this book. Readers wishing to go deeper are encouraged to explore especially the references cited in Chapter 9.

8.4.2 Who: The User and Social Context

The second element of context is concerned with 1) who the user of the mobile device is and, possibly, 2) the social context in which the user is situated. Acquiring the identity of the mobile user is usually dependent on information the user has stored in the mobile device and/or on privacy permissions that the user has set to control access to various account information. For example, on Android devices a user can create a local profile (see code example below), or personal information can be imported via an account, such as Gmail, Facebook, or Twitter. Most major web services offer an SDK for Android that allows privacy-controlled access to profile information (e.g., Facebook SDK for Android [36]).

```
public HashMap<String, String> getProfileInfo() {
   HashMap<String,String> profile = new HashMap<String,String>();
   ContentResolver cr = getContentResolver();
   Cursor cursor = cr.query(
      // Specify the URI of the "Profile" contact.
      Uri.withAppendedPath(
      ContactsContract.Profile.CONTENT_URI,
      ContactsContract.Contacts.Data.CONTENT_DIRECTORY),
      Query.PROJECTION, null, null, null);
   cursor.moveToFirst();

   // Loop through each cursor row, appending key/value to HashMap
   while (!cursor.isAfterLast()) {
      String key = null;
      // Determine key for HashMap based on MIMETYPE column in
      // cursor row. Example shown for name field. Other fields
      // left for reader to implement as desired.
      if (cursor.getString(Query.MIMETYPE).contains("/name")){
```

```
            key = "email";
        }
        if (key != null){
            profile.put(key, cursor.getString(Query.DATA));
        }
        cursor.moveToNext();
    }
    return profile;
}
```

The amount and type of information that can be obtained via these methods varies greatly based on the particular information source. For each particular user, it also depends on what information that user has entered into his or her profile and what accounts are accessible on the device. Therefore, it is important to implement fallback methods, such as manual prompts, in case required user information is not available via automated methods.

8.4.3 Where: The Location Context

As discussed above, location is an important element of context. Since much of this book is concerned with describing mobile-based techniques to obtain a user's position, we will not repeat such descriptions here. In the Android platform, there are various techniques used to obtain geographic coordinates (i.e., latitude, longitude, and altitude) that hide much of the underlying complexity used to calculate them. Most importantly, the `LocationManager` class provides methods to obtain location (i.e., position) updates from either a GNSS[23] or a network-based provider. Table 8.1 below highlights the most important methods from this class.

It is important to note that location can be expressed in many different forms, such as in geodetic coordinates (see Section 2.1) or semantic descriptions like "at home." From the perspective of context awareness, it is often such semantic descriptions that provide the most readily useful information. In this form, the context is rather self-evident, whereas the contextual meaning of geodetic coordinates (e.g., latitude and longitude) must be interpreted.

This interpretation could be relatively straightforward, such as querying a database of points of interest (recall this topic from Chapter 7), or it could be complex, such as analyzing the location history of a particular user to infer the relevance of the location. In either case, it is clear that the relevant information is not the geodetic coordinates themselves but rather the significance of the location which the coordinates reference.

---

²³ As noted in Chapter 1, GNSS is the preferred term when referring to satellite-based navigation systems in general. In Android, there are classes and methods using the acronym GPS when most modern mobile devices have multi-GNSS enabled receivers.

There is one other location-related topic that has not been covered elsewhere in this book, that of *microlocation sensing*. This refers to the use of proximity-based positioning (recall Section 2.2.6) to detect that a mobile device is within a very small-scale region (e.g., centimeter and meter scale), such as near a user's desk or near a particular display in a store. This concept has also been referred to as "hyper local." As the size of the region becomes smaller, oftentimes the context becomes clearer. For example, if a mobile device has localized itself as lying on the nightstand next to the user's bed, then the context is likely related to sleep. Therefore, this topic is of particular relevance to context awareness.

Table 8.1

List of Useful Methods from `android.location.LocationManager` Class

Property	*Description*
addGpsStatusListener	Adds a GNSS status listener, which is powerful way to get low-level updates from the GNSS receiver.
addNmeaListener	Adds an NMEA status listener, which can receive NMEA sentences as they are output from the GNSS receiver.
addProximityAlert	Adds an event listener that is triggered when the device enters or exits a given circular region (supplied as latitude/longitude/radius).
getAllProviders	Returns a list of all available location providers.
getGpsStatus	Returns a GpsStatus object, containing the current state of the GNSS engine. Used in conjunction with `addGpsStatusListener`.
getLastKnownLocation	Returns a location object, containing the last known location from a specified provider (i.e., GNSS or network).
requestLocationUpdates	Used to set up events, which are triggered when a location update is available from the specified provider (i.e., GNSS or network).

There are several technologies related to microlocation sensing. For example, in late 2013 Apple starting using devices called iBeacons in its stores to detect at a fine scale where customers are located. These beacons use Bluetooth low-energy RF signals to detect when mobile devices are nearby and communicate with them accordingly. Qualcomm also offers devices called Gimbal Proximity Beacons, which operate in a similar fashion but are designed to interact with the Gimbal platform (mentioned in Section 8.3). Bluetooth low-energy is not the only signal that is used for microlocation sensing. Samsung also sells an add-on product for its mobile devices, called TecTiles, which are programmable NFC stickers. They can be programmed to interact with a mobile device whenever it is within NFC range (a few centimeters), allowing for extremely fine-scale localization. Imagine the potential applications if such NFC were deployed more widely. NFC tags could be integrated into such items as pockets in clothing and in handbags, laptops and desktop computers (e.g., for identification purposes), and cashier or payment systems. The possibilities are endless!

8.4.4 When: The Time and Date Context

The next element of context is trivial to obtain from a mobile device's OS. For example, in Android you can get the device's system time with two lines of code:

```
SimpleDateFormat sdf = new SimpleDateFormat(
    "MM/dd/yyyy HH:mm:ss");
String dateTime = sdf.format(System.currentTimeMillis());
```

A string representing the date and time (in local time zone) will be stored in the variable `dateTime` for use in your application. The difficult part can come in figuring out how to use this information.

A date and time coordinate must be interpreted in much the same way a location coordinate is interpreted. For example, 6:00 a.m. for one person might represent a normal weekday wake-up time, whereas for another person it might represent a precious "Do not disturb" sleep time. Similarly, 5:00 p.m. on December 21 might represent an entirely different context than 5:00 p.m. on June 21 (especially at higher latitudes). Last, significant dates, such as holidays and personal events (birthdays, anniversaries, etc.), must be inferred using some additional information source. These difficulties make inferring the context from a raw time and date measurement a nontrivial matter.

There are various techniques available for inferring the contextual relevance of time and date information. The most obvious choice is by utilizing the built-in calendar application with which most (if not all) mobile devices are equipped. The drawback of this approach is that it relies on information supplied by the user, which may not be complete or up-to-date. In most cases, however, it is the easiest of the available options to implement.

An alternative approach would be to analyze the activity history of a user to infer the likelihood that he or she would perform a certain activity, such as commuting to or from work, during certain times, days of the week, or dates of the calendar year. This approach, however, may not yield detailed enough information for some applications, and it is considerably harder to implement.

8.4.5 Why: The Motivational Context

The fifth element of context—the *motivational context*—is perhaps the most elusive of them all. It involves analyzing one or more other contextual elements in order to infer a user's intentions. The techniques used to perform such inferences are largely dependent on the specific contextual domain or application where the inferences will be utilized. We can already see several minor examples of motivational context being applied in commercial devices, although clearly we are in the early stages of development in this regard. For example, Motorola's Moto X smartphone uses its rich set of sensors and specialized processors, as well as the features of Google Now, to infer the needs and desires of its users. This includes sensing when the phone is picked up (in order to turn on the screen and display notifications) and sensing certain hand gestures (in order to activate the camera application).

In general, techniques from machine learning and pattern recognition could be used. It is clear, however, that further research and development is needed before motivational context can be inferred and utilized on a large scale. To build up accurate models of human behavior and human intent, it may be necessary to generate large amounts of training data, perhaps using a crowd-sourcing approach. This topic will be considered further in Chapter 10.

8.4.6 In What Manner: Motion Context and Other Details of Context

The sixth element of context can rightly be judged as somewhat of a catch-all of other contextual information. It is formed from the sixth component of Hermagoras's *circumstance* and can be used to groups together additional details of a situation that do not fit into the other elements. For example, if the answer to *what* is the activity "dancing," then *in what manner* might include such concerns as the type of dance and the tempo of the dance. In particular, motion attributes fit nicely into this category. For example, one can use motion recognition techniques to detect the particular mode of motion and to describe other motion attributes such as speed, heading, and acceleration.

8.4.7 By What Means: The Context-Aware Device and the Methods of Sensing Context

The last element of context includes descriptions of the context-aware device itself, as well as the methods employed to sense the context. Given the wide range of mobile devices available today (recall Chapter 1), it is important for an application to be aware of details about the device it is operating on. This will supply the application with information about the capabilities of the device and about what other contextual information might be available to it. Some details about the state of the device can even constitute a situation or context themselves (e.g., low battery context).

Basic information about the device for use in an application, such as device name, manufacturer, model, operating system, and build version, is generally available via the OS's APIs. For example, the Android APIs make available a set of system properties in the class `android.os.Build`. See Table 8.2 for a list of some of these properties, which may be useful for describing the mobile device.

In addition, it is generally straightforward to obtain contextual information about the data communication functionalities of the mobile device, such as whether or not a call is in progress and what cellular network or WLAN the mobile device is connected to. In Android, one can use the `TelephonyManager` class to obtain information related to the cellular radio. Examples of useful methods available from this class are shown in Table 8.3. Similar information about the WLAN radio of an Android device can be obtained using the `WLANManager` class. Alternatively, if one is interested in data connectivity regardless of whether it is through mobile networks, WLAN, Bluetooth, or other means, the `ConnectivityManager` provides basic information about the connected or available data networks.

Table 8.2

List of Useful System Properties (Fields) from `android.os.Build` Class

Property	*Description*
Device	Codename given to the device (e.g., maguro)
Brand	Brand-customized version of the operating system (e.g., Verizon)
CPU	The name of the instruction set (CPU type + ABI convention) of native code (e.g., armeabi-v7a)
Display	An ID representing the build version of the operating system (e.g., JRN84D)
Manufacturer	Manufacturer of the device (e.g., Samsung)
Model	Model name of the device (e.g., Galaxy Nexus)
Product	Codename given to the firmware version of device (e.g., takju)
Radio	Version of the cellular radio firmware (e.g., I9250XXKK6)
Serial	Serial number of the device (e.g., 014682070502301D)
Version.release	Version number of the device's operating system (e.g., 4.1)

Other information about the device's current state or condition, such as the current battery state, amount of memory available, and current usage of the CPU, is often also available. For example, in Android, detailed information about the battery state can be obtained from the BatteryManager class. For memory usage, use the ActivityManager. For CPU usage, there is no specialized method for obtaining overall usage, but since Android is based on Linux, one can use the `/proc/stat` file to obtain the desired information.

In addition to these classes and methods, Android offers a powerful technique to make applications aware of various actions or events triggered on the device, known as *intent messaging*. Intents are essentially intradevice messages, and by setting up *intent filters* in an application, the app can receive these messages, containing various types of information depending on the type of intent. A small sampling of intents that are triggered automatically by different system events includes: answering a phone call, pressing the "call" button, unlocking the keyguard, starting the camera application, plugging in a headset, switching airplane mode on or off, docking into a desktop or car dock, installing or removing an application, rebooting or shutting down the device, low battery warning, and low storage space warning.

Table 8.3

List of Useful Methods from `android.telephony.TelephonyManager` Class

Method	Description
getCallState()	Returns an integer representing the call state of device (e.g., CALL_STATE_RINGING = 1)
getDataState()	Returns an integer representing the data state of device (e.g., DATA_CONNECTED = 2)
getDataActivity ()	Returns an integer representing the current data activity of the device (e.g., DATA_ACTIVITY_OUT = 2)
getDeviceId()	Returns an integer that uniquely identifies the device (IMEI, MEID, or ESN) (e.g., 352563176148781)
getNetworkOperatorName()	Name of the cellular network operator (e.g., FI SONERA)
getNetworkType()	Returns an integer representing the type of cellular network currently in use (e.g., NETWORK_TYPE_UMTS = 3)
getPhoneType()	Returns an integer representing the type of radio used to transmit voice calls (e.g., PHONE_TYPE_GSM = 1)
getSimSerialNumber()	Serial number of the SIM card (e.g., 8893579030109339463)
getSubscriberId()	Unique subscriber ID for mobile customer (e.g., IMSI)
isNetworkRoaming()	Returns a Boolean value depending on network roaming state

Finally, custom intents can also be implemented, which in effect allows any software component in the device to be aware of any action of any other component, provided the appropriate intents and intent filters are specified. Indeed, intent messaging provides a powerful means to implement various types of context awareness in Android devices.

8.5 HOW TO USE CONTEXT

Now that we've seen an overview of the various elements of context, we will look at how context can actually be utilized in context-aware applications. In order to demonstrate this, we return to the fictional scenario outlined in Section 8.1.1, where Mary and her colleagues are waiting for their colleague Steve to arrive, in order to leave together for lunch.

It turns out that the colleagues had gathered for lunch after responding to an *"AreUin?"* request from Mary, a mobile app that they frequently use in their department to see who is interested in joining for lunch or for after-work social outings. Steve had answered the request with *"I'm in,"* so it is a bit odd that he hadn't shown up in the lobby yet.

Timo says to Mary, "Did you try calling Steve?"

"Yes, but he didn't answer. *Locator* says that he's in the lab. It could be though that he left his phone there because it shows as completely static for the past hour and connected to a PC. I'll run and check."

Meanwhile, Timo starts the *"RestaurantFinder"* app on his phone to look for possible restaurants. The app sorts nearby restaurants according to the preferences of those who are joining for lunch, whose identities have been sent from the *"AreUin?"* app. *RestaurantFinder* shows at the top of the list that a new Mexican restaurant opened up this week—only a 10-minute drive from their work—and that Mexican food ranks highly among the group's preferences.

"There's a new Mexican place that opened up on Turuntie. You guys interested?" asks Timo.

"Sure," says Anindya, as the others also nod approvingly. At that moment, Mary returns alongside Steve.

"Sorry guys, I had my headphones on and was working at the soldering table. Lost track of time," explains Steve.

"No worries, Steve. But you do know that AreUIn has a built-in reminder feature, right?" replies Timo with a wink of the eye.

"Really? I didn't know."

"Yeah, it can even connect to your smartwatch, in case you don't have your phone on you. So you can rock out to Zeppelin and forget about us…until it vibrates as a reminder," jokes Timo. "Anyways, it's good Mary found you because there's a new Mexican place nearby called Pancho's, and I already checked to see if there's a table open. Not only was the answer yes, but they just sent me a coupon. 20% off for groups of five or more! That sound OK to you, Mary?"

"Sounds great. I have a big car, so I'll drive. Can you send me the address?"

"Done. Ok, let's go!"

As our characters walk to the parking lot, Mary's smartphone has already started a navigation application to show the way to "Pancho's." Let's turn away for a moment and take a look at how context has been utilized in this story.

First, Steve's position—more precisely the position of Steve's phone—was determined by the locator app. Furthermore, Mary had some contextual information that Steve's phone had been idle for some time and connected to a

PC. This is an example of how motion patterns and information about the phone's state can be used to provide additional details beyond mere position.

In addition, this example illustrates an important point; when the smartphone is the sole sensing platform, the inference of *user context* is dependent on the smartphone being in close proximity to the user. If Steve had simply left his phone in the lab and was in another location, Mary would have been out of luck. This is one of the reasons why context awareness may shift in the future toward wearable devices, such as smartwatches. Such wearable devices can still communicate and cooperate with other mobile devices, but in addition they have considerable sensing and networking capabilities of their own. This technology is still in its infancy, but already nearly a dozen manufacturers are selling various incarnations of this idea. Wearable devices and sensors will be discussed further in Chapter 10.

Next, Timo used the *social context*, namely the identities supplied by the "*AreUIn?*" app and their preferences stored by the *RestaurantFinder* app, to select a suitable restaurant. *RestaurantFinder* also used the approximate position of Timo's phone to filter the choices, which is a typical feature in LBSs.

Last, Timo performed a context-dependent query to determine if the restaurant currently had a table available for five people. The restaurant replied in the affirmative and also appears to have engaged in what we might call *context-sensitive marketing*: based on the busyness of the newly opened restaurant and the opportunity to bring in five potentially regular customers, it offered a discount coupon. Let's return to the story to see if context was further employed during this lunch outing.

After their meals at Pancho's were served, the jovial colleagues got into a lively discussion about soccer. "Barcelona is going all the way this year. Messi is healthy now and playing incredibly well," argues Steve.

"They will go far, but my pick is Bayern Munich. The team is just as strong as last year's, and Pep Guardiola is doing a great job as manager," counters Timo.

"Hey guys, look." Mary holds up her phone to show a notification. The notification shows that the monthly department meeting starts in 15 minutes.

"Oh, man. I totally forgot. It's the third Friday of the month, isn't it?" notes Anindya.

"Yep. I'm glad I put it in my calendar. Let's get going. Waiter, check please!"

Here we see a somewhat more innovative use of context. Mary's phone had detected that she was located several kilometers from her workplace, in fact, a 10 minute drive. Since her calendar showed a meeting starting at 1:30 p.m., and it was already 1:15 p.m., it displayed a high-priority notification. Apparently the others had either not put this meeting into their calendars or had not set up such notifications. Luckily for them, Mary had done both, so they avoid being late this time.

This fictional example provides an example of how context can be utilized in context-aware devices, namely smart mobile devices. From our introductory examination of context awareness in this chapter, it should now be fairly straightforward to envision how these features could be implemented. Admittedly, this example is not as "futuristic" and impressive as the *ubiquitous computing* scenario that Weiser depicted in 1991. Our intention was to demonstrate the concepts in this chapter and to provide a bridge to more advanced context awareness capabilities researchers and developers can implement with the current generation of mobile devices.

This example also demonstrates that context awareness, at least at the present level of technology, does not completely replace human intelligence and initiative. It can, however, go a long way in unburdening humans from various tasks, both mental and physical (in a human-computer interaction sense), freeing our human capacities for other purposes. Thus, the role of the context awareness researcher and developer should be now clearer—we must minimize the burden required to perform various tasks and open up new possibilities that would not be possible without context awareness.

8.6 SUMMARY

This chapter introduced the concept of *context* in terms of its importance to mobile geospatial computing. Since context is notoriously difficult to define, we have presented an intuitive framework for understanding the various elements of context, centered around seven questions: *what, who, where, when, why, in what manner,* and *by what means*. This in turn was used to explore the concept of *context awareness*, which we defined broadly as the quality of having knowledge of context. We briefly examined context awareness from a historical perspective, where we could see it blossoming as a research topic in the 1990s.

Next, we examined each of the elements of context in more detail, providing some examples of how contextual information within particular elements can be obtained. For several of these elements, including *what, why,* and *in what manner,* the appropriate methods vary greatly depending on the desired contextual information. In addition, these are often the most challenging elements to implement. For these reasons, further coverage of these topics is reserved for Chapter 9 on *contextual reasoning.*

Last, we used a fictional story, depicting a common workspace scenario in order to demonstrate how context awareness can be used in mobile devices. This story also highlighted several new concepts, such as *social context* and *context-sensitive marketing*, which are increasingly important in modern mobile devices.

References

[1] McCarthy, J., "Notes on formalizing context," *Proceedings of the 13th International Joint Conference on Artificial Intelligence*, pp. 555-562, San Mateo, California: Morgan Kaufmann, 1993. [Online]. Available: http://bit.ly/OEHAyO.

[2] Akman, V., and M. Surav, "Steps toward formalizing context," *AI Magazine*, Vol. 17, No. 3, 1996, pp. 55-72, available: http://citeseerx.ist.psu.edu/viewdoc/summary?doi=10.1.1.48.3474.

[3] Foltz, P. W., W. Kintsch, and T. K. Landauer, "The measurement of textual coherence with latent semantic analysis," *Discourse Processes*, Vol. 25, No. 2 & 3, pp. 285-307, 1998.

[4] Bazire, M., and P. Brézillon, "Understanding context before using it," *Modeling and Using Context: Lecture Notes in Computer Science*, pp. 113–192, A. Dey, B. Kokinov, D. Leake, and R. Turner (eds.), Berlin, Heidelberg: Springer Berlin/Heidelberg, 2005, available: http://dx.doi.org/10.1007/11508373_3.

[5] "Context," In *Merriam-Webster.com*, 2012. [Online]. Available: http://www.merriam-webster.com/dictionary/context.

[6] "Five Ws," In *Wikipedia.com*, 2012. [Online]. Available: http://en.wikipedia.org/wiki/Five_Ws.

[7] Bennett, B. S., "Hermagoras of Temnos," In *Classical Rhetorics And Rhetoricians: Critical Studies And Sources*, pp. 187-193, M. Ballif and M. G. Moran (eds.), Westport, CT: Praeger Publishers, 2005.

[8] Welch, J. R., and K. G. Salter, "A context algorithm for pattern recognition and image interpretation," *IEEE Transactions on Systems, Man and Cybernetics*, Vol. SMC-1 , No. 1, 1971, pp. 24-30.

[9] Toussaint, G. T., "The use of context in pattern recognition," *Pattern Recognition*, Vol. 10, 1978, pp. 189-204.

[10] Swain, P. H., S. B. Vardeman, and J. C. Tilton, "Contextual Classification of Multispectral Image Data," NASA Technical Report SR-PO-00443, 1980.

[11] McCarthy, J., and P. J. Hayes, "Some philosophical problems from the standpoint of artificial intelligence," *Machine Intelligence*, Vol. 4, 1969, pp. 463-502.

[12] Google, Inc., Google Now web site, [Online]. Available: http://www.google.com/landing/now/.

[13] Morishige, R. I., and J. Retelle, "Air combat and artificial intelligence," *Air Force Magazine*, October 1985, pp. 91-93.

[14] Endsley, M. R., "Design and evaluation for situation awareness enhancement," *Proceedings of the Human Factors Society—32nd Annual Meeting*, 1988, pp. 97-101.

[15] Gilson, R. D., "Special Issue Preface," *Human Factors* (Special Issue on Situation Awareness), Vol. 37, No. 1, 1995, pp. 3-4.

[16] Dey, A. K., "Context-Aware Computing," In *Ubiquitous Computing Fundamentals*, Boca Raton, FL: CRC Press, 2010, pp. 321-352.

[17] Brézillon, P., "Context in human-machine problem solving: A survey," *The Knowledge Engineering Review*, Vol. 14, No. 1, 1999, pp. 1-34.

[18] Frege, G., *The Foundations of Arithmetic, a Logico-Mathematical Enquiry into the Concept of Number,* transl. J. L. Austin, Oxford: Basil Blackwell, 1974. Originally published in German as *Die Grundlagen der Arithmetik, eine logisch mathematische Untersuchung über den Begriff der Zahl,* Breslau: Verlag von Wilhem Koebner, 1894.

[19] Frege, G., "On sense and reference," *The Philosophical Review*, Vol. 57, No. 3, 1948, pp. 209-230. Originally published in German as "Über Sinn und Bedeutung," *Zeitschrift für Philosophie und philosophische Kritik*, Vol. 100, 1892, pp. 25-50.

[20] Russell, B., "On denoting," *Mind*, New Series, Vol. 14, No. 56, 1905, pp. 479-493.

[21] Bledsoe, W. W., and I. Browning, "Pattern recognition and reading by machine," *1959 Proceedings of the Eastern Joint Computer Conference*, 1959, pp. 225-232.

[22] Miller, G. A., G. A. Heise, and W. Lichten, "The intelligibility of speech as a function of the context of the test materials," *Journal of Experimental Psychology*, Vol. 41, No. 5, 1951, pp. 329-335.

[23] Duda, R. O., and P. E. Hart "Experiments in the recognition of hand-printed text: Part II—Context analysis," 1968 *Proceedings of the Fall Joint Computer Conference*, 1968, pp. 1139-1149.

[24] Porzel, R., *Contextual Computing: Models and Applications*, Heidelberg: Springer, 2011.

[25] McCarthy, J., "Situations, Actions, and Causal Laws," Stanford Artificial Intelligence Project Memo No. 2, 1963.

[26] McCarthy, J., "Generality in artificial intelligence," *Communications of the ACM*, Vol. 30, No. 12, 1987, pp. 1030-1035.

[27] Giunchiglia, F. "Contextual reasoning," *Epistemologia, Special Issue on "I Linguaggi e le Macchine*," Vol. XVI, pp. 345-364, 1993.

[28] Giunchiglia, F., and P. Bouquet "Introduction to contextual reasoning," *Perspectives on Cognitive Science*, Vol. 3, 1997, pp. 138-159.

[29] Weiser, M., "The computer for the 21st century," *Scientific American*, Vol. 265, No. 3, 1991, pp. 94-104.

[30] Want, R., et al., "The active badge location system," *ACM Transactions on Information Systems*, Vol. 10, No.1, 1992, pp. 91-102.

[31] Schilit, B. N., and M. M. Theimer, "Disseminating active map information to mobile hosts," *IEEE Network*, Vol. 8, No. 5, 1994, pp. 22-32.

[32] Brown, P. J., "The stick-e document: a framework for creating context-aware applications," *Electronic Publishing*, Vol. 8, No. 2/3, 1995.

[33] Abowd, G. D., et al., "Cyberguide: a mobile context-aware tour guide," *Wireless Networks.*, Vol. 3, No. 5, 1997, pp. 421-433.

[34] Dey, A. K., G. D. Abowd, and A. Wood, "CyberDesk: a framework for providing self-integrating context-aware services," *Knowledge-Based Systems*, Vol. 11, 1998, pp. 3–13.

[35] Moran, T. P., and P. Dourish (eds.), "Special Issue on Context-Aware Computing," *Human-Computer Interaction*, Vol. 16, Nos. 2-3, 2001.

[36] Facebook, Inc., "Android SDK Reference," [Online]. Available: http://bit.ly/PoIpbX.

Chapter 9

Contextual Reasoning

In the last chapter we introduced the concepts of context and contextual information. In this chapter, we look more deeply at how smartphones can obtain and process this type of information. There are a number of terms that have been suggested to encompass the idea of processing contextual information, such as contextual thinking [1] and contextual intelligence [2]. We have chosen *contextual reasoning* [3], but all of these terms are more or less equivalent. We will present several examples of the techniques used for contextual reasoning, though we will not attempt to be exhaustive in our coverage. To aid those wishing to go deeper in this topic, we will point out ample references where additional information can be found.

9.1 WHAT IS CONTEXTUAL REASONING?

We begin with a formal definition. *Contextual reasoning is the process of forming higher level inferences about context from lower level information.* "Higher" and "lower" are relative terms in this definition, where the higher levels are more general and abstract, and the lower levels are more specific details that individually don't provide the "big picture" of a context. In colloquial terms, it is the process of seeing the forest through the individual trees.

Figure 9.1 illustrates this process conceptually. Known as the "context pyramid," this figure divides the contextual reasoning process into six different abstraction levels, although in practice there may be fewer or more levels, depending on the application. The first level represents the raw data received from sensors, whether we speak of hardware sensors like accelerometers and gyroscopes, or "soft sensors" like a GNSS receiver or application code that detects various states of the smartphone (e.g., low battery warning). In the next level of abstraction, physical parameters are derived from the raw sensor data, such as the speed that the user is moving or decibel level recorded by a microphone. In the third level (which may be optional in some applications), various features or patterns may be extracted from the lower levels, such as combining successive parameters to calculate a moving average or variance of the parameters.

The next three levels represent the "heart" of contextual reasoning. We move from the realm of numeric data to more semantic representations of context. In the first such level, simple contextual descriptors are inferred from the lower-level data, such as the motion pattern of the user (e.g., standing and walking), or a semantic representation of the user's position (e.g., at work). In the next level, several simple contextual descriptors may be combined to infer a higher-level context, such as the user's current activity. Finally, all of the available contextual information, including information from external sources, may be combined to a final level, which (if successful) can be described as a *rich context*.[24] Ideally this should be expressed in natural language in a form that approaches prose. This is the final result of asking and answering the questions from the journalistic framework described in Chapter 8.

9.2 A HYPOTHETICAL EXAMPLE

To make these concepts more concrete and to provide a geospatial computing example, consider the following low-level information presented in Table 9.1 from a hypothetical example.

A logical high-level inference that might be made from this information is, "Mary is walking to work." This could probably be considered an activity-level descriptor (level 5) in terms of the context pyramid as shown in Figure 9.1. If additional contextual details were to be added, such as the time of day, weather conditions, or any interesting events or details that occurred in the recent past, then it might rise to the level of rich context. As one can imagine, the exact boundaries between these levels are blurry, and the context pyramid should be considered only as a tool for understanding the process of contextual reasoning.

Note that probably several intermediate steps were required in the process of inferring, "Mary is walking to work." For example, we might have calculated that at Mary's current position, she is 500m away from the FGI and that her heading indicates she is going toward the street that leads to FGI.

Clearly humans make inferences like "Mary is walking to work" all the time. It is one of the main functions of our brain to perform this kind of reasoning. Sometimes there are a few intermediate steps performed in the subconscious, and the inference is performed instantaneously. Other times, there may be many intermediate steps, and the inference might require careful analysis of all the available information. It might even require hypothesizing and probing for additional information to confirm a hypothesis.

---

[24] In using the adjective "rich," we invoke the concept from AI of a *rich object*, which is defined as "an object which cannot be completely described or represented but about which assertions can be made" [33].

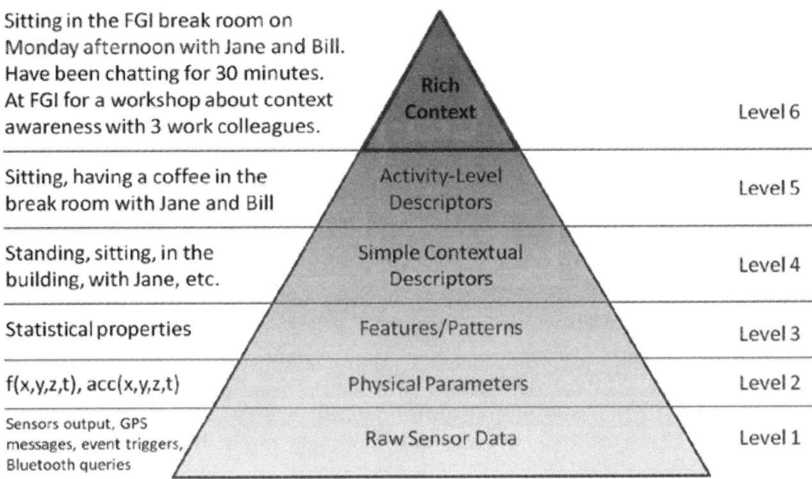

Figure 9.1 The context pyramid.

Table 9.1

Example of Contextual Information, Its Source, and Corresponding Level

Information	Source	Level
Mary is located at 60.161 N, 24.542 E.	GPS	Level 2
Mary has a heading of 170 degrees.	GPS	Level 2
Mary is walking.	Motion classifier	Level 4
Mary works at FGI.	Personal profile	External
FGI is located at 60.161 N, 24.546 E.	Significant location DB	External

The same is true for computer-based contextual reasoning. Sometimes there is enough available information and the inference is simple enough to be performed in realtime without any human interaction. In other cases, it can be extremely complex (probabilistic in nature), and require significant processing power. It may even require additional input from the user (i.e., human-computer interaction).

Consider the following additional information related to the above example:

It is 6 p.m. on a Sunday evening.
Mary normally works from 9 a.m.–5 p.m. on weekdays.

From this additional information, a computer-based contextual reasoning program might output that there is a 60% chance Mary is walking to work and a 40% chance she is walking somewhere else. Due to this ambiguity, the program queries the system for additional information:

Mary is walking with her dog.
Mary lives at 60.10 N, 25.00 E.

After this additional information and a few intermediate calculations, the program outputs:

Mary is walking her dog, near her workplace and headed away from her home.

We can see from this example that there are different inferences that can be made from different subsets of low-level information. Furthermore, we see that, in general, the richer the set of information we have to formulate our inferences, the better chance we have of generating a correct one.

Finally, observe that contextual reasoning can also be thought of as a process of *encapsulation* of information. For many applications, it is not interesting to the user nor directly useful for the application that "FGI is located at 60.10 N, 25.10 E" or (worse) that "the standard deviation in Mary's speed for the past 5 seconds is 0.58 m/s." What the application (or user) needs to know is "what is Mary doing," "who is she with," "in what manner is this activity playing out," and similar higher level questions. Thus, contextual reasoning can be alternatively thought of as the process of transforming the multitude of available information into a more concise and useful contextual result, where the definition of "useful" is highly dependent on the desired application.

9.3 WHAT ARE THE METHODS OF CONTEXTUAL REASONING?

The primary tool for making contextual inference is *machine learning*, also known as *statistical learning* or *pattern recognition*.[25] Machine learning is deeply rooted in probability theory, decision theory, and information theory. Unfortunately, we can't cover all of these topics in this chapter, but we refer those readers who are not familiar with these topics to introductory texts on machine learning, where these topics are adequately covered. In truth, it could take years of careful study to deeply understand the state-of-the-art in machine learning and contextual reasoning. Fortunately, however, there are several simple, basic techniques that work well for many applications. This section is intended to give a gentle

---

[25] Historically, pattern recognition grew out of engineering disciplines, whereas machine learning is the term adopted by most computer scientists [5]. For obvious reasons, statistical learning is the term favored by statisticians. O'Connor humorously noted that at Stanford University there are two almost identical courses offered: "machine learning" by the computer science department and "statistical learning" taught by the statistics department [6].

introduction to some of the common techniques of machine learning that are applied to contextual reasoning.

9.3.1 Introduction to Machine Learning

According to Mitchell et al., "machine learning research seeks to develop computer systems that automatically improve their performance through experience" [4]. In the domain of contextual reasoning, this means that, given the tools of machine learning, the more the system is used, the better it should be able to reason about the context the user is in. The word *learning* is appropriate because the system adaptively learns how to recognize different contexts that it encounters, in the same way that a small child (or even adult) learns to recognize different situations based on his or her experiences. In both cases (machine and human), there is a set of inputs or stimuli that are used to reason that a certain situation or context exists. The process of learning (which happens naturally and almost automatically for humans) is figuring out the optimal way to transform a set of inputs into the correct output—in our case, the context.

In general, the output of a machine learning algorithm can be of two types: 1) one or more continuous variables represented by real numbers, 2) a discrete class, which is normally represented by a textual descriptor (e.g., green, male, good, and happy) or an integer or binary code, from a finite set of classes. When the output is of the continuous type, the machine learning task is known as *regression*. When it is of the discrete type, it is known as *classification*. Since context is primarily of a discrete nature, we will focus mostly on classification in this chapter.

In classification, we can understand the learning problem as a process of learning a function, $f(\cdot)$, that maps a vector of inputs $x = (x_1, x_2,...x_n)$ to the correct output $y \in Y=\{y_1, y_2,...y_m\}$. This is expressed symbolically as:

$$f : x \rightarrow y \tag{9.1}$$

Note that the function need not be a deterministic one. It may consist of probability distributions of the input and output variables, describing a stochastic relationship between them. Ultimately, however, a classifier chooses a single class based on the function's inputs and using a particular selection criterion, such as *maximum likelihood*.

Machine learning techniques mostly fall into one of two categories based on how the function f is learned: 1) *supervised learning* and 2) *unsupervised learning*. In supervised learning, a set of "training data" is used. The training data is a limited set of input data for which the correct or optimal output is known [i.e., $s = (x, y)$]. This is sometimes known as labeled data because in most cases a human has manually labeled the dataset with the correct output. Therefore, it is usually laborious and costly to obtain such data. On the other hand, in unsupervised learning no training data are used, and the goal is to recognize implicit patterns in the data. There is, in fact, a third category of machine learning, known as *semisupervised learning*, in which both labeled and unlabeled data are used. The aim in semisupervised learning is to combine techniques from supervised and unsupervised learning and use all available data (labeled and unlabeled) to achieve a better learning result. In this book, we will mainly focus on supervised learning techniques, as they are the most commonly used in context recognition.

Most of the techniques in machine learning have been developed to deal with data samples that are independent from one another. For example, in a machine learning algorithm to classify credit card applications as "high-risk" or "low-risk," it is reasonable to assume that one credit card application does not influence the risk level of another credit card application. In contextual reasoning, however, our data samples mostly (if not exclusively) come from *time-series data*, where the assumption of independence does not strictly hold.

We can understand this intuitively from the fact that what a person is doing at one moment usually affects what he or she does at the next moment. In other words, a data sample at any given time epoch is dependent on the values of the data at prior time epochs, especially those epochs immediately prior to the given epoch. To a certain extent, we can choose to ignore these dependencies, in order to use a machine learning model with a more simple structure, but the best performance will be achieved when these time dependencies are taken into account. The subset of techniques from statistics and machine learning that work with this kind of data is known as *time-series analysis* [7] or *sequential machine learning* [8, 9].

There are certainly many such techniques to choose from, each having relative advantages and disadvantages. Table 9.2 presents a condensed list of techniques from machine learning that could be applied to contextual reasoning. It is not exhaustive, but provides a broad overview of the most important machine learning techniques for contextual reasoning. A subset of these techniques will be discussed at an introductory level in the sections below.

It is tempting when presented with a long list of machine learning techniques to want to find and focus on the "best" one, applying it indiscriminately in any given domain. The *no free lunch theorem for supervised learning* states, however, that no single machine learning algorithm performs better than any other across all problems. Thus, it is unavoidable that some process of comparison or analysis must be carried out for any particular domain before the most appropriate machine learning technique is chosen.

9.3.2 Naïve Bayes' Classifiers

Before we turn to examples of sequential machine learning techniques, we start with an example of a simple machine learning technique, in order to motivate more complex concepts later on. This simple yet surprisingly effective classifier is called the *naïve Bayes' classifier*. It falls within a larger set of powerful graphical probability models, called *Bayesian networks*, but takes advantage of a few simplifying assumptions.

Table 9.2

List of Machine Learning Techniques and Associated References

Supervised Learning	*References*
Used with sliding window method:	
Naïve Bayes'	[10-11] [13]
Bayesian networks	[24]
Decision trees	[25] [42]
LDA/QDA	[34-36] [38-40]
Logistic regression	[40-42]
SVMs/kernel machines	[26-28] [34] [41]
Neural networks	[40] [42]
Used directly on time-series data:	
HMMs	[18-19] [23] [35] [43]
Conditional random fields	[29-32] [34] [43]
Unsupervised Learning	*References*
Clustering	[37]
Principal component analysis	[36]
Self-organizing maps	[44-46]

First, we assume that all data samples are *independent and identically distributed* (*iid*). In other words, we ignore the time dependence between the data samples, and we assume that the set of processes producing the data are stable throughout the dataset (i.e., stationary processes). Second, although the data samples may be multivariate, we assume that all of the measured variables are conditionally independent from one another given the class. In many (if not most) applications, these assumptions are not strictly true, but the time dependencies and the possible dependencies between the measured variables are simply ignored in order to reduce the problem to a more tractable solution, hence the name "naïve."

Despite these simplifications, naïve Bayes' classifiers have been shown to perform well in a number of machine learning domains, including document classification [10], image classification [11], and activity recognition [12, 13]. We will use them as a starting point in our discussion of contextual reasoning because they illustrate a number of fundamental concepts before we move on to more complex machine learning techniques.

The basic structure of a naïve Bayes' classifier is shown graphically in Figure 9.2, where the unshaded node represents a class set $C_k = \{c_1, c_2, ..., c_n\}$ and the shaded nodes represent a d-dimensional input vector $x = (x_1, x_2, ..., x_d)$. The arrows represent the dependencies between the classes and the input vector. The lack of arrows between the variables in the input vector indicates that they are conditionally independent given the class.

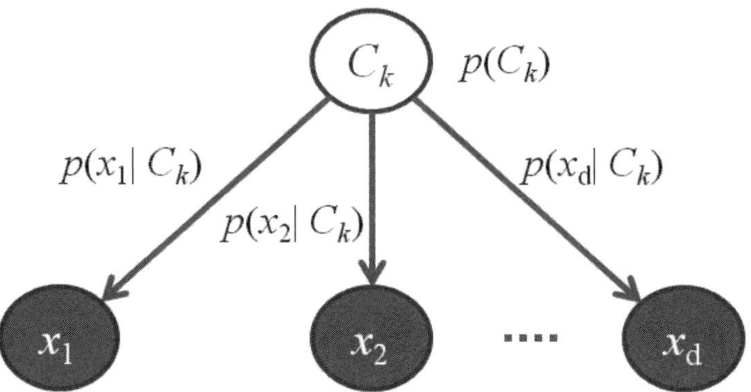

Figure 9.2 Structure of the naïve Bayes' classifier.

We assume that, in general, the class cannot be observed directly, whereas the input vector can be; therefore, we would like to classify a dataset into classes, according to this input vector. In other words, we want to know the posterior probability distribution of C_k given x, which according to Bayes' theorem is:

$$p(C_k \mid x) = \frac{p(C_k)p(x \mid C_k)}{p(x)} = \frac{p(C_k)p(x \mid C_k)}{\sum_k p(x \mid C_k)p(C_k)} \qquad (9.2)$$

In order to evaluate the likelihood $p(x|C_k)$, we would ordinarily need to compute the product:

$$p(x_1 \mid C_k)\prod_{i=2}^{d} p(x_i \mid x_1,...,x_{i-1},C_k),$$

expressing the conditional dependence among the input variables within each class, which may be very difficult if the number of input dimensions is large. By assuming that the input variables are conditionally independent given the class, the product reduces to:

$$p(\mathbf{x}|C_k) = \prod_{i=1}^{d} p(x_i|C_k) \qquad (9.3)$$

The most common way to evaluate $p(x|C_k)$ is to use a training dataset, where the values are of x are labeled with the correct class (i.e., supervised learning). If x is composed of discrete variables (e.g., with multinomial distribution), $p(x_i|C_k)$ can be estimated as the frequency that x_i takes on a particular value for a particular class; stated another way:

$$p\left(x_i = x_j \middle| C_k = C_m\right) = \frac{\text{count}(x_i = x_j, C_k = C_m)}{\text{count}(C_k = C_m)} \tag{9.4}$$

If x is composed of continuous variables, then other methods must be used to estimate this class conditional density, such as a histogram, kernel estimator, or a k-nearest-neighbor approach. Alternatively, one can use parametric methods, especially if certain assumptions about the distribution can be made (e.g., that it is Gaussian). For more details on parametric and nonparametric estimation methods, see [14].

The prior probabilities of the classes, $p(C_k)$, can also be estimated using frequency counts:

$$p(C_k = C_m) = \frac{\text{count}(C_k = C_m)}{N} \tag{9.5}$$

where N is the total number of training samples. Thus, $p(C_k = C_m)$ is simply the fraction of the training samples that are labeled with class C_m. In some cases, however, it may be that the training data disproportionately represent certain classes compared to the real-world case. Therefore, one can use domain knowledge to adjust the class priors appropriately. For example, in some applications, it may be appropriate to use equal probabilities for each class. Just be sure that the probabilities over all classes sum to one.

Note that in (9.2), the denominator will be the same for all classes. Since our ultimate goal is to choose the most likely class [i.e., the largest posterior probability, $p(C_k|x)$], it is not necessary to actually compute the value of the denominator. Given data sample x_j, one can find the largest value of $p(C_k)p(x_j|C_k)$ and classify the sample into the respective class given by the arguments.

9.3.3 Hidden Markov Model (HMM)–Based Classifiers

In the next section, we turn to a slightly more complex but more powerful machine learning technique that uses HMMs. Generally speaking, HMM-based classifiers will perform better than simple naïve Bayes' classifiers for time-series data because they are able to model the time dependency in the data. For this reason, HMMs are one of the most popular machine learning techniques for contextual reasoning.

9.3.3.1 Markov Chains and the Markov Property

First, we introduce the concept of a *Markov chain*, a mathematical system used to model stochastic processes that exhibit a *Markov property*, which will be described shortly. These processes are usually interpreted as belonging to a system that can be, at any given time, in one state from a set of possible states, S. In this chapter, we will consider only discrete state processes with a finite set of possible states, S = {S₁, S₂,...Sₘ}. The system transitions from one state to another in a stochastic manner, meaning that the transitions can be analyzed using probabilities. Such a system is represented graphically in Figure 9.3(a), where the probabilities governing the evolution of states {1, 2, 3} are shown as arrows between the nodes in the graph or as loops back onto themselves.

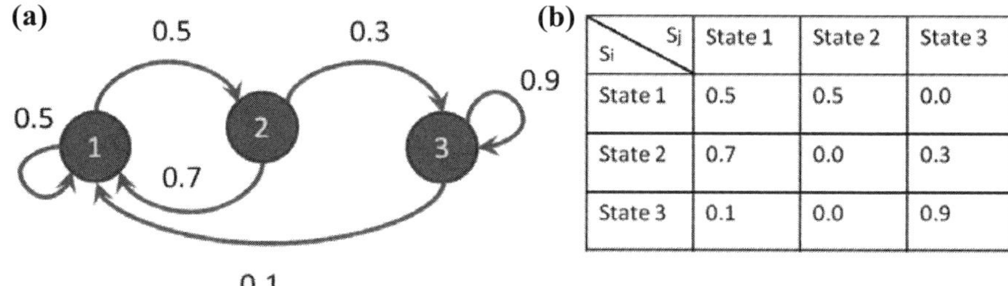

Sⱼ \ Sᵢ	State 1	State 2	State 3
State 1	0.5	0.5	0.0
State 2	0.7	0.0	0.3
State 3	0.1	0.0	0.9

Figure 9.3 A Markov chain and associated transition probability matrix.

For example, if the system is in state 1 at time t, it has a 50% probability of transitioning at time $t+1$ to state 2 and a 50% probability of remaining in state 1. Note that the lack of an arrow or loop indicates a zero probability of that particular transition. The full set of transition probabilities can also be expressed as an M x M matrix, where M is the number of states, as shown in Figure 9.3(b) for this example Markov chain. These correspond to conditional probabilities, where element ij is the conditional probability of transitioning from the ith state to the jth state at any epoch t.

When a Markov chain can be fully specified in this manner, it is said to be a first-order Markov process. The two-dimensional structure of the transition probabilities ensures that all of the information available to predict the next state of the process is contained in the present state. In probability terms, this means that the conditional probability of the system being in state s_j at time $t+1$, given knowledge that the system is in state s_i at time t, is equal to the conditional probability of the same outcome at $t + 1$, given the same knowledge of the system's state at time t plus information of its state at time $t - 1$ and earlier; i.e. stated another way:

$$P\big(S(t + 1) = s_j \big| S(t) = s_i\big) =$$
$$P\big(S(t + 1) = s_j \big| S(t) = s_i, S(t - 1) = s_k, ..., S(1) = s_l\big) \qquad (9.6)$$

When (9.6) holds for a system, we say it has the *first-order Markov property*. The second-order Markov property is when the state at $t + 1$ depends on the current state and the state at $t - 1$. This, however, would require a three-dimensional transition probability matrix ($M \times M \times M$) to be fully specified. In general, it is also possible for a system to have an nth-order Markov property, which would require an n-dimensional transition probability matrix. Only first-order Markov processes will be considered further in this chapter.

In other words, a (first-order) Markov process' exact "path" to a particular state does not provide any more useful information than the current state of the system. An example of this kind of process would be a chess game, where in order to assess the possible next move, the current configuration of the board is just as useful as the exact sequence of moves in the game up to the current configuration.

9.3.3.2 Hidden States and Observable Signals

There are two differences between an ordinary Markov model and a HMM. The first is that in HMMs, the state of the system at any given moment is not directly observable (i.e., is hidden). Second, for each transition of states, the system "emits" an observable signal, which is stochastically correlated with the unobservable state. All that one can "know" about the system's state is contained in the observed signal (which may have several components), and one can make probabilistic inferences about the state based on the correlation between the states and the observations.

As an example, consider again a chess game. Suppose now that an observer is far away and cannot make out the board but can see the players' faces and gestures and also knows something about their characters. For example, based on previous experience, this observer knows that player 1 usually smiles broadly after making a good move and often groans when she realizes she made a bad move. Player 2 is much more calm and relaxed during the game but almost always smiles the moment he realizes he has checkmate.

This could possibly be modeled using a HMM where the states are {*game even, player 1 winning, player 2 winning, player 1 wins, player 2 wins*}. The observed signals are the smiles, gestures, and sounds of the players, but the actual state is "hidden" because the board is far away. Another variant of this example would be where the observer just doesn't know the rules of chess, so is unable to assess the state, except at the very beginning and very end.[26]

HMMs can either be discrete or continuous, depending on the nature of the observed signal. For simplicity we will consider only discrete HMMs in this chapter, but many approaches have been developed to handle continuous observation signals, for example, by describing the signal parametrically. See [19] for examples.

Figure 9.4(a) shows a graphical representation of a HMM, similar to that of a Markov chain but adding the observation signal. In this example, there are only

---

[26] This would be an example of a partially HMM. See [15] for details.

two possible states {1, 2}, and the lighter shading indicates they are not directly observable. In this particular example, the transition probabilities constrain the system to only occasionally change states, since most of the probability for each state (90%) is concentrated in the loops. These transition probabilities are shown as before in Figure 9.4(b). In addition, there are two possible output signals Y = {X, O}. In state 1, these are expressed with equal probability, whereas state 2 has a tendency to express more X's than O's (80% vs. 20%). These "emission" probabilities are shown in Figure 9.4(c). In general, for discrete HMMs the emission probabilities are given by an MxN matrix, where M is the number of states and N is the number of observation symbols.[27]

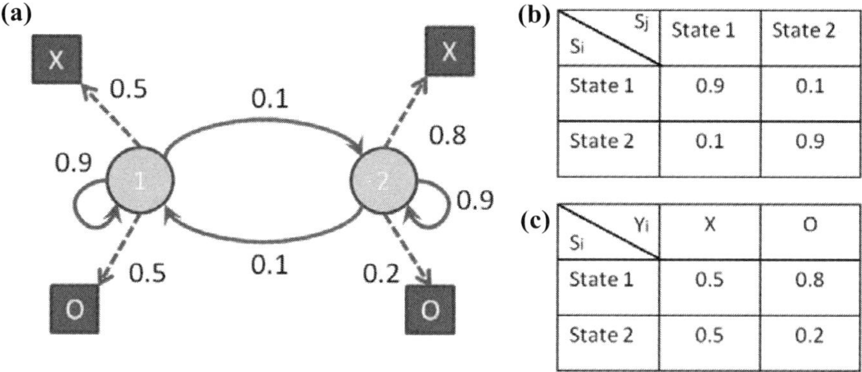

(b)

S_i \ S_j	State 1	State 2
State 1	0.9	0.1
State 2	0.1	0.9

(c)

S_i \ Y_i	X	O
State 1	0.5	0.8
State 2	0.5	0.2

Figure 9.4 A Markov chain and associated transition probability matrix.

9.3.3.3 Machine Learning Tasks Using HMMs

Now that we have laid out the basic concepts of HMM, let us explore how they are used in machine learning and contextual reasoning. From here forward it will be important to use clearly defined mathematical notation for HMMs in order to keep the text concise. We adopt the notations listed in Table 9.3, which have already been partially used above.

There are primarily four machine learning tasks that are commonly performed with HMMs, and each task has its own standard algorithm. These tasks are described as follows:

---

[27] For continuous HMMs, the emission probabilities might be expressed by a parametric model (e.g., a Gaussian with different parameters for each state).

- **Supervised model learning.** Given a sequence of observations Y_1^T that are labeled with the true sequence of states S_1^T, estimate the model parameters λ.[28]

- **Unsupervised model learning.** Given a sequence of observations Y_1^T and an initial guess of the model parameters λ_0, find the set of parameters λ_n that optimally matches the observations.

- **Online recognition.** Given a model λ and a sequence of observations Y_1^t, find the most likely state s_i at time t. Repeat for each epoch, $t=1,\ldots,T$ to obtain the sequence of states S_1^T. This is also known as sequential processing.

- **Offline recognition.** Given a model λ and a sequence of observations Y_1^T, find the most likely sequence of states S_1^T [i.e., maximum $\Pr(S_1^T|Y_1^T,\lambda)$]. This is also known as batch processing.

Table 9.3

Mathematical Notations of the HMMs

Notation	*Description*
T	The number of elements in a sequence of observations.
Y_1^T	A sequence of observations, indexed from 1 to T.
M	The number of possible states in the model.
$S = \{s_1, s_2, \ldots, s_M\}$	The set of possible states.
$S(t)$	The state at time t.
S_1^T	A sequence of states, indexed from 1 to T.
N	The number of possible observation symbols.
$Y = \{y_1, y_2, \ldots, y_N\}$	The set of possible observation symbols.
$Y(t)$	The observation symbol at time t.
$a_{ij} = \Pr\left(S(t+1) = s_j|S(t) = s_i\right)$	The conditional probability that the system will be in state s_j at time $t+1$, given the state s_i at time t. Also called the transition probability from state i to j.
$b_k = \Pr\left(Y(t) = y_k|S(t) = s_j\right)$	The probability that state s_j will emit observation symbol y_k. Also called the emission probability for y_k and s_j.
$B = \{b_j\,(k)\}$ for $j = 1, \ldots, M$ and $k = 1, \ldots, N$	The set of emission probabilities for all observation and symbols and states, also called the emission probability matrix.
$\pi = \Pr\{S(1) = s_j\}$ for $j = 1, \ldots, M$	The set of initial state probabilities for each state s_j.
$\lambda = \{A, B, \pi\}$	The complete parameter set of the model.

There are two subtle but important differences between task 3 and task 4. First, task 4 is an offline task, so the algorithms used to achieve it can use the full

---

[28] Note that we have just broken the assumption about the state of the system not being directly observable. This is true only for the model learning phase (i.e., training). During the recognition phase, we still assume the states are hidden.

set of observations Y_1^T at any time in its operation, whereas task 3 can only use past observations. Second, task 3 operates at each epoch individually and will produce an optimal choice of state at that particular epoch (given the available observations), whereas task 4 chooses the optimal sequence S_1^T, which has the highest probability of producing the observations Y_1^T. In other words, they have two distinct choices of optimality criterion, which can sometimes lead to differing results.

Supervised Model Learning

This is the most straightforward of the four tasks. First, to estimate the transition probability matrix A, we simply need to count up the occurrences of each transition type $i{\mapsto}j$ (including where j=i) for all i,j={1,...,M}. The transition probabilities are then estimated as:

$$a_{ij} = \frac{\text{count}(S(t)=s_i, S(t+1)=s_j)}{\text{count}(S(t)=s_i)} \tag{9.7}$$

This will give us M^2 transition probabilities, allowing us to fully specify the matrix A. In practice, however, some of these transition probabilities will be zero. This *could* be because the probability of that particular transition is actually zero, or it could be simply because the transition is not represented in the limited training sequence S_1^T. Therefore, it is good practice to manually inspect the matrix A and use one's domain knowledge to determine if some possible transitions are not represented in the dataset when they are indeed possible. One approach to rectify this problem is to add a *pseudocount* to each such transition. The value of the pseudocount can either be one (Laplace's rule) or can be set to some other positive integer value based on domain knowledge.

Next, to estimate the emission probability matrix B, we count for each state the number of times each observation symbol has been expressed. The emission probability is then estimated as:

$$b_j(k) = \frac{\text{count}(Y(t)=y_k, S(t)=s_j)}{\text{count}(S(t)=s_j)} \tag{9.8}$$

Again we can use Laplace's rule or some other pseudocount value, based on domain knowledge, in order to correct any probabilities that would otherwise be unreasonably set to zero.

Last, we must estimate the initial state probabilities that should be specified in π. In practice, the methods for setting this parameter are application-dependent. The simplest approach is to set it to an equal value for each state that sums to one over all states (i.e., $1/M$). If for some particular application, it is more likely that the HMM is initialized in some particular state or states, the values for these states can be adjusted according to domain knowledge (maintaining that the probabilities over all the states sum to one). Another possibility is when the initial state of an HMM is known with great certainty, the value for that state can be set to one and all other initial state probabilities set to zero. Note that no matter how parameter π is specified, its importance diminishes as t increases, so even the simple $1/M$ approach may be adequate.

Unsupervised Model Learning

In this task, we would like to estimate the parameters of the HMM based on an unlabeled observation sequence Y_1^T. In order to do so, we need an initial guess of the parameters, λ_0. The standard algorithm for optimizing the parameters based on this initial guess and an observation sequence is called the *Baum-Welch algorithm* or the *forward-backward algorithm*. A detailed treatment of this algorithm (which includes several variants) is beyond the scope of this chapter, but we only mention here that it is an expectation–maximization (EM) procedure, where an improved estimate of the model parameters is given after each iteration, until converging to a local maximum of $Pr(Y_1^T|\lambda)$. The result varies based on the initial guess of the parameters, so one approach is to run the algorithm a number of times with randomly generated initial guesses (subject to reasonable constraints). This won't necessarily yield a globally optimized set of parameters, but no finite time approach to obtain a global maximum of $Pr(Y_1^T|\lambda)$ is known. More details about the Baum-Welch algorithm can be found in [14] or [19].

Online Recognition

Given a HMM defined by parameters λ and a history of observations Y_1^T, the goal of this task is to determine the most likely state s_i at time t. In other words, we need to find the state that gives the maximum probability $Pr(S(t)|Y(t),\lambda)$ from the set of all possible states, also known as the maximum a posteriori (MAP) estimate of S(t):

$$\hat{S}(t)_{MAP} \equiv \arg\max_i \Pr\left(S(t) = s_i | Y(t), \lambda\right) \tag{9.9}$$

In order to evaluate this probability, we use an identity called the Chapman–Kolmogorov equation and the Markov property to obtain a prediction equation:

$$p(S(t) = s_j \mid Y(t-1)) = \sum_{j=1}^{M}\sum_{i=1}^{M} a_{ij}\, p(S(t-1) = s_i \mid Y(t-1))\delta(s_j - s_i) \tag{9.10}$$

where $\delta(s_j - s_i)$ is the Dirac delta measure that is equal to zero for $s_i \neq s_j$ and one for $s_i = s_j$ [16]. Then, given a new observation $Y(t)$, we use Bayes' rule to obtain the updated equation:

$$p(S(t) \mid Y(t) = k) = \frac{b_j(k)p(S(t) = s_j \mid Y(t-1))}{\sum_{i=1}^{N} b_i(k)p(S(t) = s_i \mid Y(t-1))} \qquad (9.11)$$

These two equations form the basis of the prediction and update steps of this recognition algorithm, respectively. At each time epoch, the prediction density $p(S(t) = s_j \mid Y(t-1))$ is calculated and then updated using the new observation and (9.11). This technique is often known as Bayesian optimal filtering or recursive Bayesian estimation. In some references, it has been called the grid-based method, presumably because the transition probabilities are represented as a grid (i.e., matrix) [16].

Offline Recognition

The goal of this task is to find the most likely sequence of states S_1^T, given a model λ and a sequence of observations Y_1^T. Since this task is performed after *all* of the observations in the sequence are available, there is more information available to perform it compared to the online recognition task. Several different algorithms have been used in order to perform this type of task, but the most commonly used is the Viterbi algorithm [14].

The Viterbi algorithm defines two matrices, $\delta(i,t)$ and $\psi(i,t)$. $\delta(i,t)$ represents "best path" probabilities for traveling particular paths in the sequence of states. For example, $\delta(3,4)$ is the probability of reaching state 3 at time 4 using the best or most likely path to get there. $\psi(i,t)$, on the other hand, represents "back pointers" to the states that produce the most probable paths. For example, if $\psi(3,4) = 2$, this means that, in order for the HMM to be in state 3 at time 4 with maximum probability $\delta(3,4)$, it must be in state 2 at time 3. Using the same notation $\delta(i,T)$ represents the probability of being in state i at the end of the sequence (i.e., time T).

The algorithm starts at time epoch one and then recursively calculates the successive values in $\delta(i,t)$ and $\psi(i,t)$. When the values of $\delta(i,T)$ have been calculated, the algorithm simply chooses the value of i that gives the maximum $\delta(i,T)$ and then backtracks through the sequence using values of $\psi(i,t)$ to determine the most likely path.

Formally, the Viterbi algorithm is implemented in four steps: 1) initialization, 2) recursion, 3) termination, and 4) path backtracking [17], described as follows.

1) Initialization: The values of $\delta(i,t)$ and $\psi(i,t)$ for time $t=1$ are set:

$$\begin{aligned} \delta(i,1) &= \pi_i b_j(y_1) \\ \psi(i,1) &= 0 \end{aligned} \qquad 1 \leq i \leq M \qquad (9.12)$$

Note that $\psi(i,1)$ has no clear meaning, since it specifies a state of the system prior to the first epoch. By convention it is set to zero, but it is never actually used in the algorithm.

2) **Recursion:** The values of $\delta(j,t)$ and $\psi(j,t)$ for times $t=2,...,T$ are recursively calculated:

$$\begin{aligned} \delta(j,t) &= \max_{1\le i\le M}\left[\delta(i,t-1)a_{ij}\right]b_j(Y(t)) \\ \psi(j,t) &= \arg\max_{1\le i\le M}\left[\delta(i,t-1)a_{ij}\right] \end{aligned} \qquad \left\{\begin{array}{l} 2\le t\le T \\ \\ 1\le j\le M \end{array}\right. \qquad (9.13)$$

Note that we have changed the state index variable used in δ and ψ from i to j. This is done in order to remain consistent with the notation for a_{ij}, representing the probability of transitioning from state i to state j. Therefore, here i represents the state at time $t-1$ and j represents the state at time t.

3) **Termination:** Select the maximum probability, P_T, among the values of $\delta(i,T)$ and the corresponding state, $S(T)$, according to the following criteria:

$$\begin{aligned} P_T &= \max_{1\le i\le M}\left[\delta(i,T)\right] \\ S(T) &= \arg\max_{1\le i\le M}\left[\delta(i,T)\right] \end{aligned} \qquad (9.14)$$

4) **Path backtracking:** Determine the corresponding most likely path (state sequence) using the back pointers, $\psi(i,t)$, and the following backward recursion:

$$S(t) = \psi(S(t+1),t+1) \quad t = T-1, T-2,...,1 \qquad (9.15)$$

9.3.3.4 Concluding Remarks about HMMs

This concludes our coverage of HMMs in this chapter. For readers not well-versed in probability theory (especially concepts such as Markov chains), these sections may have been challenging to digest. Fortunately, implementations of HMMs and the related algorithms for working with them are readily available in many programming languages (e.g., Java, C, and Matlab). In order to use them correctly, it is necessary to understand the concepts outlined here at a basic level. We have not addressed in this chapter various implementation details, such as techniques to avoid numeric underflow or variations of the algorithms to reduce memory requirements. Therefore, particular implementations may differ somewhat from what was presented here. More detailed coverage of HMMs can be found in [5, 18, 19].

9.3.4 The Sliding Window Method

As mentioned in Section 9.2.1, most of the techniques in machine learning have been developed to deal with *independent and identically distributed* (*iid*) data (HMMs are an exception to this). Such methods, in their original form, do not exploit the time dependence inherent in sequential data. A simple but effective technique to benefit from the wide array of available machine learning techniques yet also exploit this time dependency is the *sliding window method,* illustrated in Figure 9.5 [20]. This method effectively encapsulates the (local) time dependence of the data into a new data structure, a "sliding window," allowing the time-series data to be used in the same way as one would use iid data. As a result, one can use any of the classical (iid) machine learning techniques to classify samples of this sliding window.

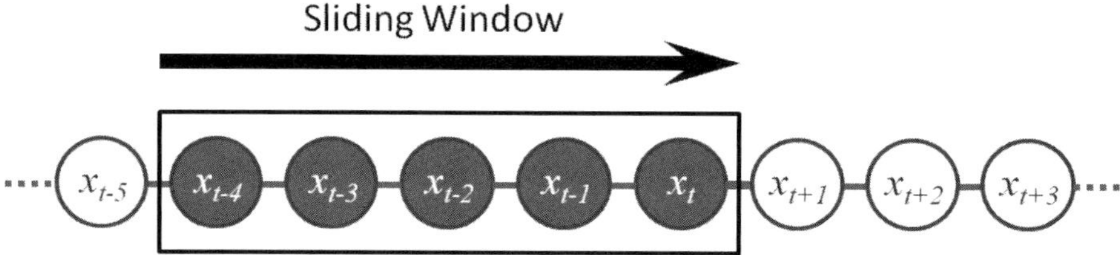

Figure 9.5 The sliding window method.

The best way to elucidate the sliding window method is by example. Suppose we have the following one-dimensional (unitless) data sequence:

$$X_1^{10} = (7,10,3,9,6,5,8,10,5,3)$$

First, a window width, w, must be chosen. The best choice of window width is dependent on the application and determines how much of the local time dependence is preserved in the data. Large windows will capture more time dependence but will also result in more computationally complex learning and classification. Therefore, the best choice is a trade-off between classification performance and computational time. In our example, we will choose a window size of five epochs.

Next, the windows are composed, which will result in a new data structure, H_t. This structure will have $w*d$ dimensions, where d is the original dimensionality of the data. In our case, H_t will be five-dimensional. There are two slightly different approaches to composing the windows, depending on how one wants to handle epochs 1 to $w - 1$, where there is not enough data to "fill" the window. The first approach is keep the number of samples in H_t the same as that of the original data and to use null values where necessary. For example, the first three windows composed from X_1^{10} would be:

$$H_1 = (\text{null}, \text{null}, \text{null}, \text{null}, 7)$$
$$H_2 = (\text{null}, \text{null}, \text{null}, 7, 10)$$
$$H_3 = (\text{null}, \text{null}, 7, 10, 3)$$

The other approach is simply to reduce the *length* of H_t by $w - 1$. Thus, epoch five from our original dataset becomes epoch one of H_t, and the first three windows become:

$$H_1 = (7, 10, 3, 9, 6)$$
$$H_2 = (10, 3, 9, 6, 5)$$
$$H_3 = (3, 9, 6, 5, 8)$$

In either case, as the window "slides" to the right in time, a new value from the sequence X_1^{10} comes in at the last (rightmost) position in H_t and an existing value (or null) slides out of the window with the values in between also being shifted accordingly.

After all the windows in H_t have been composed, the ordering of H_t is no longer important—it can now be regarded as a set instead of a sequence. In other words H_1, H_2, H_3, etc., can be shuffled like a deck of cards into any order. So long as the individual elements within each window are not rearranged, the local time dependence is preserved.

This is beneficial for many reasons. For example, in machine learning, data are usually separated into different groups for training the classifier and later for testing its performance. Now these groups can be chosen by randomly selecting samples from the set H_t, thus ensuring there is no selection bias in these groupings.

9.3.5 Bayesian Networks

Bayesian networks are directed graphical models, where nodes and directed links (i.e., edges) between the nodes represent conditional dependencies between variables (or sets of variables). As mentioned in Section 9.2.2, naïve Bayes' classifiers are a special case from the general framework of Bayesian networks. We saw in Section 9.2.2 that naïve Bayes' classifiers assume the input variables in the model are independent (i.e., no interconnecting arcs between them), but since this assumption is often not correct, Bayesian networks in general can model this interdependence among input variables. In fact, HMMs can also be viewed as a specific type of Bayesian network, called a *dynamic Bayesian network*, with certain constraints placed on its structure [23].

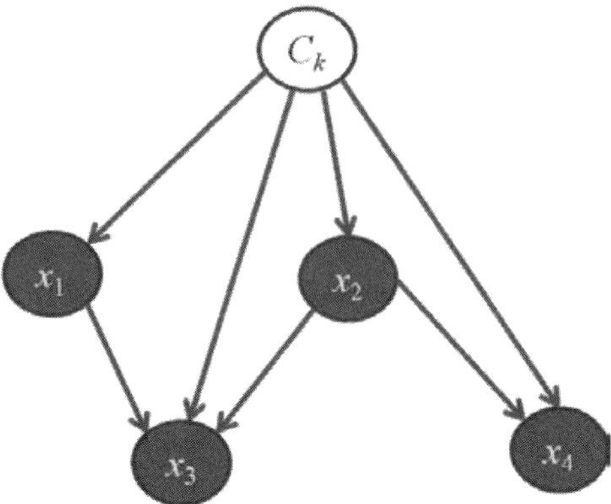

Figure 9.6 Example of a Bayesian network.

In general, Bayesian networks can be used to represent *any* joint probability distribution that consists of a product of conditional distributions. For example, the Bayesian network shown in Figure 9.6 corresponds to the joint distribution defined by the following equation:

$$p(C_k, x_1, x_2, x_3, x_4) =$$
$$p(C_k)p(x_1 \mid C_k)p(x_2 \mid C_k)p(x_3 \mid C_k, x_1, x_2)p(x_4 \mid C_k, x_2) \tag{9.16}$$

In this way, Bayesian networks are a compact and intuitive means to express the probabilistic relationships among observable variables and hidden variables, such as the class label. Unfortunately, we don't have space to cover in detail how Bayesian networks are used in machine learning, but algorithms exist both to learn the graph structure from a labeled dataset and to perform inference on a given Bayesian network given unlabeled data. For detailed treatment of these subjects, see [5] or [24].

9.3.6 Decision Trees

Decision trees, or tree-based methods (which include several variants such as CART, ID3, and C4.5), comprise a simple but widely used machine learning technique that functions as a hierarchical set of if-then-else statements. For example, in the decision tree shown in Figure 9.7(a), classification is performed starting from the top node. For a particular data sample, if the condition lying between nodes 1 and 2 is satisfied (i.e., $x_1 > 5$), the classifier moves to the second node. Otherwise, it proceeds to node 3. The pattern is continued until a "leaf node" is reached (denoted in Figure 9.7 using uppercase letters), at which point a class label is assigned for that data sample.

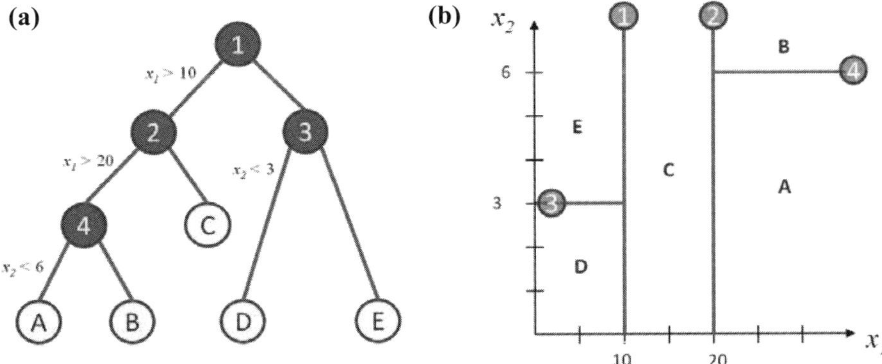

Figure 9.7 Example of a decision tree.

An alternative way to understand decision trees is that they divide an input space into distinct class regions, where each boundary corresponds to one node (i.e., "decision") from the decision tree, as shown in Figure 9.7(b). Each region can define a separate class (as in Figure 9.7), or several regions can be assigned to the same class. It is not necessary that the boundaries are orthogonal like in our example. In so-called oblique decision trees, linear combinations of variables can be used, in order to define non-orthogonal boundaries.

The overall goal with decision trees is to correctly classify as many data samples as possible, while minimizing the total number of regions (in order to avoid the problem of *overfitting*). There are two main alternatives for how to construct a decision tree. The first is to start with a single node and add one node at a time, until a threshold performance criteria is met. The other approach is to first grow a very large tree and then "prune" it (i.e., remove nodes), until a satisfactory balance is achieved between the performance of the tree and its complexity. A detailed discussion of decision trees can be found in [25].

9.3.7 Support Vector Machines (SVMs)

In recent years, SVMs have become a popular choice for building classifiers. The mathematics involved in constructing an SVM classifier are rather complex and beyond the scope of this book, but in this section we will provide sufficient detail to understand the basic concepts of SVMs.

9.3.7.1 Defining a Feature Space Using Kernel Functions

SVMs work by defining a much higher-dimensional space, called a *feature space*, mapped from the original input space. The added dimensions in the feature space are formed from transformations of one or more input variables [e.g., $\varphi: (x_i, x_j) \rightarrow$

$(x_i^2, x_i x_j, e^{x_j}, \dots)$]. Generally speaking, it is not necessary to explicitly perform a mapping from the input space to the feature space, but rather the so-called *kernel trick* is used. The kernel trick uses a *kernel function*, $k(x_i, x_j)$, which corresponds to the inner product of the feature space. Here the indices i and j reference different data samples in the input space, which are assumed to be vectors. Common kernel functions used in SVMs include:

- Homogeneous polynomials: $k(x_i, x_j) = (x_i \cdot x_j)^d$;
- Inhomogeneous polynomials: $k(x_i, x_j) = (x_i \cdot x_j + c)^d$;
- Gaussian radial basis functions: $k(x_i, x_j) = exp\left(-\gamma \|x_i - x_j\|^2\right)$, where $\gamma > 0$;
- Hyperbolic tangents: $k(x_i, x_j) = \tanh(\kappa x_i \cdot x_j + c)$.

We will demonstrate the kernel trick with a simple example, where the input space is two-dimensional ($\mathbb{R}^2$). Here we use a homogeneous polynomial kernel function where d=2. We will compute the kernel function for two samples from the input space, and for greater clarity we define these as $x = (x_1, x_2)$ and $y = (y_1, y_2)$. Note that the subscripts now reference the dimension of the input space, as opposed to earlier where they referenced the different data sample. The kernel function for x and y is computed as follows:

$$\begin{aligned} k(x, y) = (x \cdot y)^2 &= (x_1 y_1 + x_2 y_2)^2 \\ &= x_1^2 y_1^2 + 2 x_1 y_1 x_2 y_2 + x_2^2 y_2^2 \end{aligned} \tag{9.17}$$

Inspecting (9.17), we can see that the corresponding feature space has three dimensions, and the two data samples in this feature space become:

$$\varphi(x) = (x_1^2, \quad \sqrt{2} x_1 x_2, \quad x_2^2) \tag{9.18}$$

$$\varphi(y) = (y_1^2, \quad \sqrt{2} y_1 y_2, \quad y_2^2) \tag{9.19}$$

We can verify (9.18) by checking that $k(x, y) = \varphi(x) \cdot \varphi(y)$. As mentioned above, it is not necessary to explicitly map the input space to the feature space. Only the inner product, computed using the kernel function, is needed to build the classifier. In our simple example, it may seem like this is a trivial advantage, but in more complex examples, the feature space may consist of hundreds or thousands of dimensions. In fact, if a Gaussian radial basis function is used, it is usually evaluated using a Taylor series expansion. Therefore, the corresponding feature space would have infinite dimensions. Fortunately, only the inner product is needed, which can be truncated to desired precision by ignoring higher-order terms.

9.3.7.2 Maximum-Margin Hyperplanes

The main advantage of using a high-dimension space is that the data samples are more easily separated by *hyperplanes* (planes in spaces of arbitrary dimension), allowing linear classifiers to be defined in that hyperspace. These hyperplanes can be considered as similar to the boundaries between the class regions defined in decision trees (recall previous section). They are defined by maximizing the distance between the nearest data sample and the hyperplane, the so-called *maximum margin.* Such data samples lying nearest to the hyperplanes are called *support vectors* because their position supports the definition of the decision boundaries.

This concept is notionally depicted in Figure 9.8. The left side shows a two-dimensional space with data from two classes that are not linearly separable. On the right side, the data are remapped to a higher-dimensional space via a kernel function and plotted in only two selected dimensions (where the linear separation can be seen). The support vectors are the larger-sized data points lying closest to the solid red line, and this line constrains the orientation of the hyperplane in the high-dimensional space.

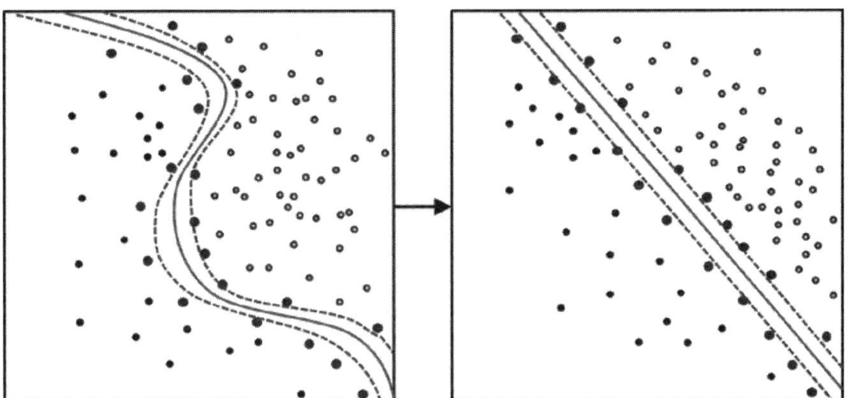

Figure 9.8 SVMs remap data into a higher-dimensional space, where the classes are linearly separable.

9.3.7.3 Concluding Remarks about SVMs

If one is interested in using SVMs to perform classification, it is not necessary to know the mathematical details about how to construct a maximum-margin hyperplane. Several open-source libraries exist for constructing SVMs and for using them to perform classification (e.g., [26, 27]). For most users of SVMs, it is probably sufficient to have a notional idea about how they function and perhaps an understanding of how different kernel functions operate. Users can experiment with different kernel functions applied to their own datasets. The performance of the resulting SVM classifier will vary, depending on the dataset and the chosen kernel function. Because the kernel functions contain one or more free parameters, *parameter tuning* is often performed to optimize the performance. More detailed coverage of SVMs, including techniques for parameter tuning, can be found in [5, 14, 28].

9.4 SUMMARY

In this chapter, we defined contextual reasoning as *"the process of forming higher level inferences about context from lower level information."* We presented this process in the framework of the context pyramid, which aims to aid understanding of contextual reasoning at the conceptual level. We also provided a hypothetical example of this process within the domain of geospatial computing in smartphones.

Next we presented the primary method of contextual reasoning, namely machine learning. We presented three machine learning techniques in detail, including the naïve Bayes' classifier, HMMs, and the sliding window method. Finally, we gave a brief overview of several other commonly used machine learning techniques, including Bayesian networks, decision trees, and SVMs. Detailed coverage of these techniques can be found in the cited references.

In conclusion, by combining the wide variety of sensor data and geospatial information available in smartphones with state-of-the-art techniques of machine learning, smartphones can be made to "reason" about the context in which their users are living. This reasoning allows smartphones to become context-aware, enabling a wide range of context-aware applications and services.

References

[1]   Puccio, G. J., M. C. Murdock, and M. Mance, *Creative Leadership: Skills That Drive Change*, San Diego, CA: Sage, 2007.

[2]   Hurdus, J. G., and D. W. Hong, "Behavioral programming with hierarchy and parallelism in the DARPA urban challenge and RoboCup," *Proceedings of IEEE International Conference on Multisensor Fusion and Integration for Intelligent Systems*, 20-22 Aug. 2008, 2008, pp. 503–509.

[3]   Giunchiglia, F., "Contextual reasoning," *Epistemologia*, Special Issue on "I Linguaggi e le Macchine," Vol. XVI, 1993, pp. 345-364.

[4]   Mitchell, T., et al., "Machine Learning," *Annual Review of Computer Science,* Vol. 4, 1990, pp. 417-433.

[5]   Bishop, C. M., *Pattern Recognition and Machine Learning*, New York: Springer, 2006.

[6] O'Connor, B., "Statistics vs. machine learning, fight!" *AI and Social Science* (blog), http://bit.ly/VtpHEX, 2008.

[7] Shumway, R. H., and D. S. Stoffer, *Time Series Analysis and Its Applications: With R Examples*, New York: Springer, 2011.

[8] Kingsbury, N., D. B. H. Tayy, and M. Palaniswamiz, "Multi-scale kernel methods for classification," *Proceedings of the 2005 IEEE Workshop on Machine Learning for Signal Processing*, 2005, pp. 43-48.

[9] Visatemongkolchai, A., and H. Zhang, "Building probabilistic motion models for SLAM," *Proceedings of the 2007 IEEE International Conference on Robotics and Biomimetics*, 2007, pp.1629-1634.

[10] Ting, S. L., W. H. Ip, and A. H. C. Tsang, "Is naïve Bayes a good classifier for document classification?" *International Journal of Software Engineering and Its Applications*, Vol. 5, No. 3, 2011, pp. 37-46.

[11] Boiman, O., E. Shechtman, and M. Irani, "In defense of nearest-neighbor based image classification," *IEEE Conference on Computer Vision and Pattern Recognition*, 2008, pp. 1-8.

[12] Tapia, E. M., S. S. Intille, and K. Larson, "Activity recognition in the home setting using simple and ubiquitous sensors," *Pervasive Computing: Second International Conference, PERVASIVE 2004*, pp. 158-175, A. Ferscha and F. Mattern (eds.), Berlin: Springer, 2004.

[13] Bancroft, J. B., D. Garrett, and G. Lachapelle, "Activity and environment classification using foot mounted navigation sensors," *2012 International Conference on Indoor Positioning and Indoor Navigation*, 2012.

[14] Alpaydin, E., *Introduction to Machine Learning, Second Edition*, Cambridge, MA: The MIT Press, 2010.

[15] Bordes, L., and P. Vandekerkhove, "Statistical inference for partially hidden Markov models," *Communications in Statistics—Theory and Methods*, Vol. 34, No. 5, 2005, pp. 1081-1104.

[16] Ristic, B., S. Arulampalm, and N. Gordon, *Beyond the Kalman Filter: Particle Filters for Tracking Applications*, Norwood, MA: Artech House, 2004.

[17] Liu, J., "Hybrid Positioning with Smart Phones," In *Ubiquitous Positioning and Mobile Location-Based Services in Smart Phones*, pp. 159-193, R. Chen (ed.), Hershey, PA: IGI-Global, 2012.

[18] Rabiner, L. R., "A tutorial on hidden Markov models and selected applications in speech recognition," *Proceedings of the IEEE*, Vol. 77, No. 2, 1989, pp. 257-286.

[19] Fraser, A. M., *Hidden Markov Models and Dynamical Systems,* Philadelphia, PA: Society for Industrial and Applied Mathematics, 2008.

[20] Dietterich, T. G., "Machine learning for sequential data: A review," *Proceedings of the Joint IAPR International Workshop on Structural, Syntactic, and Statistical Pattern Recognition*, 2002, pp. 15-30.

[21] Wolpert, D., "The lack of a priori distinctions between learning algorithms," *Neural Computation*, Vol. 8, No. 7, 1996, pp. 1341-1390.

[22] Schaffer, C., "A conservation law for generalization performance," *Proceedings of the Eleventh International Conference on Machine Learning*, 1994, pp. 259-265.

[23] Ghahramani, Z., "An introduction to hidden Markov models and Bayesian networks," *International Journal of Pattern Recognition and Artificial Intelligence*, Vol. 15, No. 1, 2001, pp. 9-42.

[24] Friedman, N., D. Geiger, and M. Goldszmidt, "Bayesian network classifiers," *Machine Learning*, Vol. 29, 1997, pp. 131-163.

[25] Sutton, C. D., "Classification and Regression Trees, Bagging, and Boosting," *Handbook of Statistics*, Vol. 24, 2005, 303-329.

[26] Chang, C. C., and C. J. Lin, *LIBSVM—A Library for Support Vector Machines*, http://bit.ly/VkqyrS, 2012.

[27] Katholieke Universiteit Leuven, *Least Squares—Support Vector Machines*, http://bit.ly/VkurwO, 2012.

[28] Steinwart, I., and A. Christmann, *Support Vector Machines*, New York: Springer, 2008.

[29] Sutton, C., and A. McCallum, "An Introduction to Conditional Random Fields," *Foundations and Trends in Machine Learning*, Vol. 4, No. 4, 2011, 267–373.

[30] Lafferty, J. D., A. McCallum, F. C. N. Pereira, "Conditional random fields: probabilistic models for segmenting and labeling sequence data," *Proceedings of the Eighteenth International Conference on Machine Learning*, 2001, pp. 282-289.

[31] Klinger, R., and K. Tomanek, *Classical Probabilistic Models and Conditional Random Fields*, Algorithm Engineering Report TR07-2-013, Technical University of Dortmund, 2007.

[32] Van Kasteren, T., A. Noulas, G. Englebienne and B. Kröse, "Accurate Activity Recognition in a Home Setting," In *Proceedings of the 10th international conference on Ubiquitous computing (UbiComp '08)*, pp. 1-9, 2008.

[33] Fisher, R.A, The use of multiple measurements in taxonomic problems. *Annals of Eugenics*, 7:179–188, 1936.

[34] Abidine, M. B., "Evaluating C-SVM, CRF and LDA classification for daily activity recognition," *Multimedia Computing and Systems (ICMCS), 2012 International Conference on*, 2012, pp. 272–277.

[35] Ward, J. A., et al., "Activity recognition of assembly tasks using body-worn microphones and accelerometers," *IEEE Transactions on Pattern Analysis and Machine Intelligence*, Vol. 28, No. 10, 2006, pp. 1553-1567.

[36] Andreu, J., R. D. Baruah, and P. Angelov, "Real Time Recognition of Human Activities from Wearable Sensors by Evolving Classifiers," *2011 IEEE International Conference on Fuzzy Systems*, 2011, pp. 2786-2793.

[37] Phung, D., et al., "High accuracy context recovery using clustering mechanisms," *IEEE International Conference on Pervasive Computing and Communications, (PerCom)2009*, 2009, pp. 1-9.

[38] Aziz, O., and S. N. Robinovitch, "An analysis of the accuracy of wearable sensors for classifying the causes of falls in humans," *IEEE Transactions on Neural Systems and Rehabilitation Engineering*, Vol. 19, No. 6, 2011, pp. 670-676.

[39] Javed, J., H. Yasin, and S. F. Ali, "Human movement recognition using Euclidean Distance: A tricky approach," *Image and Signal Processing (CISP), 2010 3rd International Congress on*, 2010, pp. 317-321.

[40] Donohoo, B., et al., "Exploiting spatiotemporal and device contexts for energy-efficient mobile embedded systems," *Proceedings of the 49th Annual Design Automation Conference*, 2012, pp. 1274-1279.

[41] Krishnan, N. C., and S. Panchanathan, "Analysis of low resolution accelerometer data for continuous human activity recognition," *Acoustics, Speech and Signal Processing (ICASSP) 2008. IEEE International Conference on*, 2008, pp. 3337-3340.

[42] Kwapisz, J. R., G. M. Weiss, and S. A. Moore, "Activity recognition using cell phone accelerometers," *ACM SIGKDD Explorations Newsletter,* Vol. 12, No. 2, 2011, pp. 74-82.

[43] Vail, D. L., M. M. Veloso, and J. D. Lafferty, "Conditional random fields for activity recognition," *Proceedings of the 6th International Joint Conference on Autonomous Agents and Multiagent Systems*, 2007, pp. 1331-1338.

[44] Kohonen, T., "The self-organizing map," *Proceedings of the IEEE* 78, no. 9, 1990: 1464-1480.

[45] Suzuki, S., et al., "Activity recognition for children using self-organizing map," *RO-MAN, 2012 IEEE*, 2012, pp. 653-658.

[46] Huang, W., and J. Wu, "Human action recognition based on self organizing map," *International Conference on Acoustics, Speech, and Signal Processing (ICASSP)*, 2010, pp. 2130-2133.

Chapter 10

Future Directions in Mobile Geospatial Computing

So far, we have addressed the topics of mobile positioning indoors and outdoors using different positioning technologies; mobile GIS and LBS applications; and context awareness in mobile devices. As mobile technologies are currently undergoing rapid development, the geospatial computing capabilities in mobile devices will likewise become enhanced. Each evolution of mobile technologies will lead to a new era of mobile geospatial applications. In the 1990s, we were amazed about the possibility of making phone calls and sending text messages using a small mobile device. Most of us could not imagine during that time that a few decades later we would be very easily capable of zooming, panning, and rotating a map with our finger gestures on the touch screen of a smartphone. It is hard for us to predict what kinds of geospatial computing capabilities mobile devices will have in about 15–20 years because mobile technologies have been developing at an extremely rapid pace. Nevertheless, we will provide a few future examples, based on our knowledge, of directions in which mobile geospatial computing will likely develop in the coming decades. These directions include 1) three-dimensional visualization of geospatial data in mobile devices, 2) advanced mobile sensing and contextual thinking, 3) geospatial computing using wearable sensors, and 4) widespread use of cloud-based geospatial computing for mobile devices.

Each of these technological pathways is a natural extension from currently existing technologies. Three-dimensional visualization of geospatial objects is an enhancement of the current two-dimensional visualizations of maps and images. Mobile sensing already takes place in modern mobile devices, but in the future its scope will enlarge from the current focus on ubiquitous positioning indoors and outdoors to include recognition of different motions and activities of mobile users. Applications of contextual thinking will improve from their current limited scope into that of a highly capable "thinking engine" in future smart mobile devices. Geospatial computing using wearable sensors will extend from the current portfolio of a few devices and a limited user base into widespread adoption of a large variety of wearable form factors. Last, cloud-based mobile geospatial computing, which is already commonplace but only used in a limited number of applications, will become a widely used architecture, increasing the capability of mobile geospatial applications in terms of processing power, storage capacity, and ubiquitous accessibility.

10.1 THREE-DIMENSIONAL VISUALIZATION OF GEOSPATIAL DATA IN MOBILE DEVICES

Most modern smartphones are already capable of visualizing three-dimensional geospatial objects [1–3]. Figure 10.1 shows a screenshot of an Android application that visualizes the three-dimensional model of a small city in Finland called Espoo [1]. Users can use the direction keys to navigate along the walking path and to change the view angles of the three-dimensional scenes. This three-dimensional visualization runs rather smoothly in current smartphones. The major challenge for visualizing three-dimensional geospatial objects is not related to the capability of the hardware components in mobile devices but rather is related to the limited coverage of three-dimensional models of the Earth. New geospatial data acquisition technologies such as light detection and ranging (LiDAR) have been developed for acquiring three-dimensional geospatial data [2]. The Google Tango project (http://www.google.com/atap/projecttango/) is the first initiative of utilizing a mobile device to generate the three-dimensional model of the space around the mobile user by collecting more than a quarter million three-dimensional measurements per second. Software such as Esri's three-dimensional city Engine [4] is now also available for creating three-dimensional city models. The number of three-dimensional city models is now increasing at a steady pace. We can expect that more and more three-dimensional mobile GIS and LBS applications will be available in the coming years when we will have sufficient coverage of three-dimensional models of the Earth and especially of city areas.

10.2 ADVANCED MOBILE SENSING AND CONTEXTUAL THINKING

Advanced mobile sensing extends the application scope of the smartphone built-in sensors from positioning indoors/outdoors to activity and motion pattern recognition. The built-in sensors in smartphones are now becoming critical mobile sensing components to sense our activities at scales ranging from personal to group and community [5, 6]. This sensing system enables the development of various innovative applications related to transportation [7], environmental impact and exposure monitoring [8], and well-being [9]. Current research suggests that built-in sensors are capable of sensing motion patterns [10], mobility contexts [11], and activities [12, 13] at a success rate of more than 90%. Most of the current solutions and algorithms, however, are based on supervised learning, which requires a training phase with a large amount of samples in order to derive a robust and accurate model. For most applications, only a limited number of data samples are available for training the models. Therefore, most of these models have limitations. For example, the model may only be valid for a particular application or a particular application domain.

Figure 10.1 Visualization of a three-dimensional city model in a smartphone.

Nevertheless, it is now possible to train models in a cloud computing system with a large number of samples based on crowdsourcing. Therefore, there is a strong potential for developing a versatile "thinking engine" running in future mobile devices. The basis of this thinking engine would be a detailed and accurate model of different kinds of human behavior, in terms of the sensor data patterns that these behaviors elicit.

An analogy can be seen in leading search engines, such as Google. These search engines would not perform very well if they could not draw upon a comprehensive "model" (i.e., index and link structure) of billions of web pages. Likewise, a mobile "thinking engine" requires an elaborate model of human behavior. If such a thinking engine would someday work as well as the Google search engine does now, there is no doubt that our future mobile devices would understand many of our intents and intervene on our behalf. For example, a mobile device will be able to answer automatically any incoming calls while you are driving with a voice message like "Sorry, I am currently driving—will call you back in 10 minutes." The thinking engine will understand your current activity of driving and estimate the driving time of 10 minutes to your destination.

The above example is a simple one. Although there are many design aspects that should be carefully considered, it can be implemented already in current smartphones. For example, it is not difficult to identify the activity of "driving" using GPS. There are a lot of other activities in our daily life, however, that are difficult to classify. The implementation of such a versatile and robust thinking engine is still a challenging task.

10.3 GEOSPATIAL COMPUTING USING WEARABLE SENSORS

Similar to smartphones, smart wearable devices such as *Google Glass* have a lot of mobile computing hardware components and built-in sensors that can be adopted for mobile geospatial computing. For *Google Glass*, these hardware components and built-in sensors include an optical head-mounted display with a resolution of 640x360 pixels, digital compass, accelerometers, gyroscopes, proximity sensor, a 5-megapixel camera that is capable of supporting 720-p video recording, radio interfaces for WLAN 802.11b and Bluetooth 4.0, 1 GB RAM, and 16-GB local storage capacity [14]. Despite all these capabilities, the device weighs only 50 grams. It runs an operating system of Android 4.0.3 or higher with a set of application development interfaces called Mirror APIs. One can use natural language to command the device (e.g., asking the device to take a picture by saying "glass, take a picture"). Unlike handheld smartphones, the user experience of wearable devices is totally different. Mobile users will interact with wearable devices in different ways than what we are used to with the current generation of mobile devices, such as smartphones and tablets.

For example, *Google Glass* will be a very good mobile device for crowdsourcing geospatial data because of its head-mounted display, head-mounted camera, and capability of using natural language to command the device. It is a hands-free solution for capturing georeferenced photos, using the built-in positioning sensors to record the viewing angle (in addition to a smartphone's GNSS receiver). In addition, it is a good solution for virtual reality applications because of the head-mounted display.

Google Glass is not the only currently existing wearable mobile device. In September 2013, Samsung released an Android-based "smartwatch," called Galaxy Gear. In terms of computing power, it features an 800-MHz processor, 4 GB of persistent memory, and 512 MB of RAM. As far as sensors, it has a 1.9-megapixel camera, two noise-canceling microphones, accelerometers, gyroscopes, and a Bluetooth low-energy module for communication. At the time of this writing, other companies selling devices in the smartwatch category include Motorola, Sony, Qualcomm, Suunto, and Pebble. Apple is also now developing a wearable device called *iWatch*. In fact, there is now an official specification for *iWatch*. Given Apple's history of revolutionizing the mobile device industry, it will be very interesting to see what role the *iWatch* will play in this unfolding story.

In any case, the *smart wearable device* is now a new member of the family of mobile devices. New geospatial computing applications will be available as a result of this new technology.

10.4 CLOUD-BASED MOBILE GEOSPATIAL COMPUTING

During the last decade, cloud computing and warehouse-scale computing have revolutionized the computing industry as a whole. The National Institute of Standards and Technology (NIST) offers the following definition of cloud computing [15]:

> *Cloud computing is a model for enabling ubiquitous, convenient, on-demand network access to a shared pool of configurable computing resources (e.g., networks, servers, storage, applications, and services) that can be rapidly provisioned and released with minimal management effort or service provider interaction.*

Cloud computing is now playing an important role in geospatial computing, due to its ability to provide ubiquitous, scalable, on-demand, and cost-effective geo-processing services [16]. The immense computational power and huge storage space are attractive for mobile geospatial computing because it often involves computationally demanding tasks and because the size of geospatial datasets are typically very large. Because most applications otherwise require continuous network access, cloud computing services can be available at all times.

In addition, cloud computing offers several advantages for mobile geospatial computing, compared to performing all computations on the mobile device, including the following:

- The possibility of carrying out heavy computation tasks in the cloud (for example, rendering of high-quality three-dimensional scenes [14], so that mobile users can obtain the results rapidly), and the possibility of tasks being performed using less of the mobile device's limited energy supply (i.e., its battery).
- The possibility of implementing cooperative applications involving a large number of mobile users connected through a cloud service.
- Access to large online geospatial datasets, such as satellite images and three-dimensional city models.

Many geospatial computing applications are already taking advantage of these benefits, but there are still relatively few service providers and their offerings are relatively limited compared to what is technically feasible. For example, Google offers many cloud-based geospatial computing services, such as the Google Maps API and Google Places API, but there would be a need to perform many other kinds of geospatial computing tasks in the cloud, such as three-dimensional visualization or motion pattern recognition. Therefore, more cloud computing companies providing a wider array of geospatial services are likely to play a role in the future.

Open-source and open-data initiatives may also play a significant factor in this area, given the success of projects such as OpenStreetMap. Especially because mobile geospatial computing often involves information subject to privacy concerns, users are likely to prefer service providers that offer transparency about their processes and policies, as opposed to tightly controlled solutions where users have little control over how their location information is used. Finding adequate solutions to such privacy concerns will also help cloud-based geospatial computing applications to become more attractive to a larger number of users.

Given the clear advantages of cloud-based geospatial computing in mobile devices, we can assert with confidence that use of this approach will increase vastly in the future. It will enable many new applications that may have seemed impossible prior to the advent of the cloud.

10.5 SUMMARY AND CONCLUSIONS

This chapter reviewed the current trends and future directions of mobile geospatial computing. We identified four key developments that have the potential to usher in a new era of mobile geospatial computing: widespread availability of three-dimensional geospatial data for visualization in mobile devices, advanced mobile sensing and contextual thinking, proliferation of wearable devices with sensors, and increased use of cloud-based geospatial computing. These developments, combined with recent and continuing improvements in mobile device and network technologies, will provide the kindling for new applications to flare up in the imaginations of developers, engineers, scientists, and entrepreneurs.

Finally, because we've come to the concluding section in this book, we offer the following overall conclusions concerning mobile geospatial computing:

- Smart mobile devices, including smartphones, tablets, and wearable devices, have proliferated in recent decades, together with technologies that have enabled mobile geospatial computing, including positioning technologies, GIS, and remote sensing imagery.
- In particular, positioning technologies have enabled positioning of mobile devices in nearly all environments. GNSS-based positioning offers the highest level of accuracy in outdoor environments where the view to the sky is mostly unobstructed. In highly urban environments and indoors, other positioning methods must be employed, including cellular-based positioning, WLAN-based positioning, and hybrid techniques.

- Due to vast improvements in the computing power of mobile devices, GIS capabilities have recently migrated from desktop environments to mobile platforms, enabling more advanced and elaborate implementations of geospatial computing in mobile devices.

- Combining one or more of the above technologies, a large variety of LBSs are now available in mobile devices. These range from emergency services such as E-911 to gaming and entertainment applications such as *SCVNGR*. The common element in all of these services and applications is that they are location-aware (i.e., they exploit the position of the mobile device, which is determined by automatic means).

- An extension of location awareness is context awareness, in which a mobile device is aware of the interrelated conditions in which it exists, such as who is using the device, what that person is doing, and why they are doing it. Various sensors and machine learning techniques can be applied to achieve context awareness in mobile devices, and the application of these techniques in mobile devices is currently in a nascent level of development.

We hope this book has given you an in-depth look into the exciting world of mobile geospatial computing and has equipped you with the knowledge necessary to contribute to its future development. It is difficult to paint a detailed picture of how mobile geospatial computing will look in the future, but there are certainly well-defined technological challenges standing clearly ahead of us. Fresh ideas and approaches may be needed to solve many of these challenges, so we hope this book will help bring new people to the field. Finally, we hope this book will help inspire novel applications of mobile geospatial computing, bringing this technology to new domains and untapped markets.

References

[1]   https://play.google.com/store/apps/details?id=com.FGI.Tapiolathree-dimentional&hl=en. Three-dimensional city model of Tapiola, Finland.

[2]   Hyyppa, J., et al., "3D City Modeling and Visualization for Smart Phone Applications," In Chen, R. (Ed.), *Ubiquitous Positioning and Mobile Location-Based Services in Smart Phones*, Pennsylvania, U.S., IGI-Global. 254-296.

[3]   Chen R., et al., "Going 3D, personal nav and LBS," *GPS World*, Vol. 21, No. 2, pp. 14-18, 2010.

[4]   http://www.esri.com/software/cityengine. Esri's 3D city Engine.

[5]   Lane N. D., et al., "A survey of mobile phone sensing," *IEEE Comm. Magazine*, Sept. 2010, pp. 140–150.

[6]   Campell, A., and T. Chaudhury, "From smart to cognitive phones," *Pervasive Computing*, July-September 2012, pp. 7-11.

[7]   Reddy, S., et al., "Using mobile phones to determine transportation modes," *ACM Transactions on Sensor Networks (TOSN)*, 6(2), Article 13, 2010.

[8]   Mun M., et al., "Peir, the personal environmental impact report, as a platform for participatory sensing systems research," *Proc. 7th ACM MobiSys*, 2009, pp. 55–68.

[9]   Lu, H., et al., "StressSense: Detecting stress in unconstrained acoustic environments using smartphones," *Proc. 14th Int'l Conf. Ubiquitous Computing (Ubicomp 2012)*, ACM, 2012.

[10] Pei, L., et al., "Using LS-SVM based motion recognition for smartphone indoor wireless positioning," *Sensors*, 12(5):6155-6175, 2012.

[11] Guinness, R., "Beyond where to how: a machine learning approach for sensing mobility contexts using smartphone sensors," *Proc. ION GNSS 2013*, Nashville, September 16-20, 2013.

[12] Pei, L., et al., "Human behavior cognition using smartphone sensors," *Sensors*, 13, 1402-1424, doi:10.3390/s130201402, 2013.

[13] Consolvo, S., et al., "Activity sensing in the wild: a field trial of Ubifit Garden," *Proc. 26th Annual ACM SIGCHI Conf. Human Factors Comp. Sys.*, 2008, pp. 1797–1806.

[14] Hardware specification of Google Glass: http://en.wikipedia.org/wiki/Google_Glass.

[15] Mell, P., and T. Grance, "The NIST definition of cloud computing(draft)." NIST Special Publication, 2011, 800: p.145.

[16] Lee, K., and S. Kang, "Mobile cloud service of geo-based image processing functions: a test iPad implementation," *Remote Sensing Letters*, 4:9, 910-919, DOI:10.1080/2150704X.2013.810821.

About the Authors

Dr. Ruizhi Chen received his Ph.D. in geodesy from University of Helsinki, M.Sc. in computer science and engineering from Helsinki University of Technology, and B.Sc. in surveying engineering from Wuhan Technical University of Surveying and Mapping. Dr. Chen is currently an endowed chair and professor at the Conrad Blucher Institute of Surveying & Science, Texas A&M University Corpus Christ. He is the editor of *Ubiquitous Positioning and Mobile Location-Based Services in Smart Phones.* He is an author/co-author of 131 scientific papers and 5 book chapters. His research results have been selected twice as cover stories in *GPS Worlds*. He has supervised 4 Ph.D. dissertations, and examined 4 Ph.D. theses from Finland, Spain, and Hong Kong. His Ph.D. students have received 7 international awards in best student paper competitions, including 3 student-winning papers in the Institute of Navigation for 2010, 2012 and 2013. Dr. Chen is the general chair of the IEEE conferences *Ubiquitous Positioning, Indoor Navigation and Location-based Services* for 2010, 2012, and 2014. He is the associate editor of *Journal of Navigation*, and a member of the editorial board of *Journal of Global Positioning Systems*. Dr. Chen is the chair of the council of principle investigators and research administrators of Texas A&M University Corpus Christi (2013–2014), and member of the Provost's Leadership Team of the same university (2013–2014). Dr. Chen was President of the International Association of Chinese Professionals in Global Positioning Systems (2008) and a board member of the Nordic Institute of Navigation (2009–2012). Dr. Chen's research interests include satellite navigation and mobile geospatial computing.

Robert E. Guinness is a researcher in the Department of Navigation and Positioning at the Finnish Geodetic Institute, where he explores contextual reasoning in mobile computing platforms. Earlier, he was a researcher at Laurea University of Applied Sciences in Espoo, Finland, where he contributed to R&D related to GNSS and unmanned aerial vehicles. Mr. Guinness holds a B.A. in physics from Washington University in St. Louis and a Master of Science in Space Studies from the International Space University in Strasbourg, France. He is currently also a doctoral candidate in the Department of Pervasive Computing at Tampere University of Technology in Finland. Prior to moving to Finland in

2010, Mr. Guinness worked as a systems engineer and mission scientist at Johnson Space Center in Houston, Texas, where he was mainly involved in the International Space Station program, as well as the short-lived Altair Lunar Lander project. Although his current research is focused on Earth-bound applications of contextual reasoning, he hopes this research will also benefit future space exploration missions, where context awareness and machine learning will play key enabling roles. When he's not busy with research and writing, Mr. Guinness enjoys outdoor activities in the Finnish nature and spending time with his wife and two sons.

Index

Recent Titles in the Artech House Mobile Communications Series

John Walker, Series Editor

3G CDMA2000 Wireless System Engineering, Samuel C. Yang

3G Multimedia Network Services, Accounting, and User Profiles, Freddy Ghys, Marcel Mampaey, Michel Smouts, and Arto Vaaraniemi

802.11 WLANs and IP Networking: Security, QoS, and Mobility, Anand R. Prasad and Neeli R. Prasad

Achieving Interoperability in Critical IT and Communications Systems, Robert I. Desourdis, Peter J. Rosamilia, Christopher P. Jacobson, James E. Sinclair, and James R. McClure

Advances in 3G Enhanced Technologies for Wireless Communications, Jiangzhou Wang and Tung-Sang Ng, editors

Advances in Mobile Information Systems, John Walker, editor

Advances in Mobile Radio Access Networks, Y. Jay Guo

Applied Satellite Navigation Using GPS, GALILEO, and Augmentation Systems, Ramjee Prasad and Marina Ruggieri

Artificial Intelligence in Wireless Communications, Thomas W. Rondeau and Charles W. Bostian

Broadband Wireless Access and Local Network: Mobile WiMax and WiFi, Byeong Gi Lee and Sunghyun Choi

CDMA for Wireless Personal Communications, Ramjee Prasad

CDMA Mobile Radio Design, John B. Groe and Lawrence E. Larson

CDMA RF System Engineering, Samuel C. Yang

CDMA Systems Capacity Engineering, Kiseon Kim and Insoo Koo

CDMA Systems Engineering Handbook, Jhong S. Lee and Leonard E. Miller

Cell Planning for Wireless Communications, Manuel F. Cátedra and Jesús Pérez-Arriaga

Cellular Communications: Worldwide Market Development, Garry A. Garrard

Cellular Mobile Systems Engineering, Saleh Faruque

Cognitive Radio Techniques: Spectrum Sensing, Interference Mitigation, and Localization, Kandeepan Sithamparanathan and Andrea Giorgetti

The Complete Wireless Communications Professional: A Guide for Engineers and Managers, William Webb

Digital Communication Systems Engineering with Software-Defined Radio, Di Pu and Alexander M. Wyglinski

EDGE for Mobile Internet, Emmanuel Seurre, Patrick Savelli, and Pierre-Jean Pietri

Emerging Public Safety Wireless Communication Systems, Robert I. Desourdis, Jr., et al.

The Future of Wireless Communications, William Webb

Geographic Information Systems Demystified, Stephen R. Galati

Geospatial Computing in Mobile Devices, Ruizhi Chen and Robert Guinness

GPRS for Mobile Internet, Emmanuel Seurre, Patrick Savelli, and Pierre-Jean Pietri

GPRS: Gateway to Third Generation Mobile Networks, Gunnar Heine and Holger Sagkob

GSM and Personal Communications Handbook, Siegmund M. Redl, Matthias K. Weber, and Malcolm W. Oliphant

GSM Networks: Protocols, Terminology, and Implementation, Gunnar Heine

GSM System Engineering, Asha Mehrotra

Handbook of Land-Mobile Radio System Coverage, Garry C. Hess

Handbook of Mobile Radio Networks, Sami Tabbane

High-Speed Wireless ATM and LANs, Benny Bing

Wireless Communications in Developing Countries: Cellular and Satellite Systems, Rachael E. Schwartz

Wireless Communications Evolution to 3G and Beyond, Saad Z. Asif

Wireless Intelligent Networking, Gerry Christensen, Paul G. Florack, and Robert Duncan

Wireless LAN Standards and Applications, Asunción Santamaría and Francisco J. López-Hernández, editors

Wireless Sensor and Ad Hoc Networks Under Diversified Network Scenarios, Subir Kumar Sarkar

Wireless Technician's Handbook, Second Edition, Andrew Miceli

For further information on these and other Artech House titles,including previously considered out-of-print books now available through our In-Print-Forever® (IPF®) program, contact:

Artech House
685 Canton Street
Norwood, MA 02062
Phone: 781-769-9750
Fax: 781-769-6334
e-mail: artech@artechhouse.com

Artech House
16 Sussex Street
London SW1V 4RW UK
Phone: +44 (0)20 7596-8750
Fax: +44 (0)20 7630-0166
e-mail: artech-uk@artechhouse.com

Find us on the World Wide Web at: www.artechhouse.com